TRIUMPH OF THE HEART

Triumph of the Heart

The Story of Statins

Jie Jack Li

UNIVERSITY PRESS

2009

OXFORD
UNIVERSITY PRESS

Oxford University Press, Inc., publishes works that further
Oxford University's objective of excellence
in research, scholarship, and education.

Oxford   New York
Auckland   Cape Town   Dar es Salaam   Hong Kong   Karachi
Kuala Lumpur   Madrid   Melbourne   Mexico City   Nairobi
New Delhi   Shanghai   Taipei   Toronto

With offices in
Argentina   Austria   Brazil   Chile   Czech Republic   France   Greece
Guatemala   Hungary   Italy   Japan   Poland   Portugal   Singapore
South Korea   Switzerland   Thailand   Turkey   Ukraine   Vietnam

Copyright © 2009 by Oxford University Press, Inc.

Published by Oxford University Press, Inc.
198 Madison Avenue, New York, New York 10016

www.oup.com

Oxford is a registered trademark of Oxford University Press

All rights reserved. No part of this publication may be reproduced,
stored in a retrieval system, or transmitted, in any form or by any means,
electronic, mechanical, photocopying, recording, or otherwise,
without the prior permission of Oxford University Press.

Library of Congress Cataloging-in-Publication Data
Li, Jie Jack.
Triumph of the heart : the story of statins / Jie Jack Li.
p. cm.
Includes bibliographical references and index.
ISBN 978-0-19-532357-3
1. Statins (Cardiovascular agents)—History. I. Title.
[DNLM: 1. Hydroxymethylglutaryl-CoA Reductase
Inhibitors—history. 2. Coronary Disease—drug therapy.
3. Drug Design. 4. Drug Industry—history.
5. History, 20th Century. 6. History, 21st Century.
QU 11.1 L693t 2009]
RM666.S714L5 2009
615'.718—dc22   2008039597

9 8 7 6 5 4 3 2 1

Printed in the United States of America
on acid-free paper

DISCLAIMER

No word in this book is to be regarded as affecting the validity of any trademark. Mention of specific companies, organizations, or drugs does not imply endorsement by the author or publisher, nor does mention of companies, organizations, or authorities in the book imply that they endorse the book. The author's statements are merely his own personal opinions, not those of his employer or the companies concerned. This book, a historical account of drug discovery, is not intended to dispense medical advice for individual health and drug-related issues. Please consult your physician and/or your pharmacist with regard to your medical needs.

Lipitor is a trademark of Pfizer Inc.; Mevacor and Zocor are trademarks of Merck & Company; Pravachol is a trademark of Bristol-Myers Squibb Company; Lescol is a trademark of Novartis; Crestor is a trademark of AstraZeneca. The trade names of drugs mentioned in this book are listed in the appendix.

To my former colleagues at Parke-Davis who contributed to the discovery and development of Lipitor

FOREWORD

Jack Li has established himself over the past several years as a prolific writer on the history and science underlying the discovery of medicinal drugs. In this work he weaves a fascinating tale of two of the great victories in twentieth-century medicine—understanding the link between cholesterol and coronary artery disease and developing the statin family of medicines, which together greatly benefit humankind.

Central to this interesting and complex story are the important but little-known achievements of many individual scientists. As you read this book, you will be introduced to these major figures behind the dramatic advances in the prevention of atherosclerosis and cardiovascular diseases: Nikolai N. Anitschkov, Akira Endo, Konrad E. Bloch, Joseph L. Goldstein, Michael S. Brown, Alfred W. Alberts, Alvin K. Willard, Faizulla G. Kathawala, Bruce D. Roth, and Roger S. Newton. Because of their work, the lives of many millions of people are being significantly extended—they will enjoy extra years of health and productive life.

In this book Jack Li illuminates many critical aspects of the statin story beyond the science. For instance, he sharply depicts the challenge facing any company involved in the discovery of new medicines. He also raises some serious questions on the future of pharmaceutical discovery, which, as he points out, is quite uncertain at this moment. That future depends more critically than ever on faith, inventiveness, and perseverance in the discovery process and the willingness of the public and governments to share in the enterprise—to share the risks, costs, and uncertainty of a very unpredictable undertaking. The only certainties are that predictions are themselves risky and that more, not less, scientific and medical research is essential to modern societies.

E. J. Corey
Harvard University

PREFACE

THE STORY OF statins is fascinating—not only because of its scientific and medical importance, but also because of the human aspects in the discovery and development of these drugs. Statins are used to treat patients with high cholesterol who are at risk for heart attack, stroke, or other coronary heart disease. Now millions of patients have benefited from statins in preventing coronary heart disease (25 million in 2005 in the United States alone). Statins, the new wonder drugs by all accounts, lower LDL cholesterol (low-density lipoprotein, the "bad" cholesterol) by inhibiting the action of a key enzyme in the liver involved in the biosynthesis of cholesterol. They are considered the gold standard therapy and are the most widely prescribed lipid-modifying drugs.

"Discoveries are made by men, not merely by minds, so that they are alive and charged with individuality."[1] In telling the story of Lipitor, Zocor, and other statins, I have tried to illustrate the human side of science as well as elements of luck, persistence, and insight that characterize major discoveries. Moreover, knowledge of the history of drug discovery is essential to an intellectual understanding of it. In preparing this book, I was torn between my scholarly and popularizing instincts. It is my sincere hope that I have achieved a certain balance of both. And I hope this book will find its way to the general public in addition to scientists in academia and the drug industry, health care professionals, and students in medicine.

I am indebted to Professor Akira Endo, the discoverer of the first statin, mevastatin, for providing his stories and photographs. I thank David Canter, Roger S. Newton, and Bruce D. Roth for taking the time to reminisce with me about the exciting days during the discovery and development of Lipitor. Nick Terrett and Ian Osterloh, the key players in the discovery and development of Viagra, graciously communicated with me about their experience.

I have incurred many debts of gratitude to my academic friends Professor E. J. Corey at Harvard University and Professor Phil S. Baran at Scripps Research Institute, who have offered much enthusiastic encouragement. Professor Baran and his students Jeremy Richter and Jonathan Lockner also kindly proofread the manuscript and provided many invaluable suggestions.

PROLOGUE

In the afternoon of that Friday, December 19, 2005, Judge Joseph J. Farnan, Jr., of the Federal District Court in Delaware handed down his judgment validating Pfizer's two patents on atorvastatin, the principal ingredient of the cholesterol-lowering drug Lipitor.[1] As a consequence, Pfizer was allowed to exclusively market in the United States the world's best-selling drug until June 2011, and it would now be illegal for the challenger, India's largest generic drug company, Ranbaxy Laboratories Ltd., to make a copycat version of Lipitor.

"Today marks a major victory for medical inventors and the patients who depend on them for important new therapies," Pfizer CEO Hank McKinnell commented.[2]

Although the New York Stock Exchange was closed when the court ruling was announced, the news sent the Pfizer common stocks soaring 11.3% to $25.14 during after-hours trading. The drug industry breathed a collective sigh of relief, knowing that innovation would be protected by U.S. patent law. The judgment also ignited a broad-based advance in other pharmaceutical industry stocks. Merck surged more than 7%, Bristol-Myers Squibb rose more than 3% at one point, and Schering-Plough climbed more than 5%.

How could the two patents for a single molecule, atorvastatin, be worth billions of dollars? To answer that question, we will need to take a look at another molecule, cholesterol, whose levels atorvastatin lowers in the human body.

CONTENTS

Foreword   ix
    E. J. Corey

1 Cholesterol   3
    The Janus-Faced Molecule   4
    One of the 10 Greatest Discoveries in Medicine   11
    The Good, the Bad, and the Ugly   13
    Hitler's Gift   16
    The Framingham Heart Study   19
    The Receptor   22

2 Genesis of Statins   27
    Early Cholesterol Drugs   27
        Nicotinic acid   28
        Resins   30
        Fibrates   32
    I Refuse to Sign This Report!   33
    Birth of Statins   36
        Endo's discovery of the first statin   37
        The fate of mevastatin   40
        Endo's fate   42
        How do statins work?   43

3 Merck's Triumph   45
    Merck   45
        Humble beginnings   45
        George W. Merck   46
        Penicillin   48
        Cortisone   51
        Streptomycin   53
        Glorious history   55

Mevacor and Zocor 56
 The poor Greek immigrant kid 56
 Prelude 58
 Discovery 59
 Halt! 61
 Triumph 63
The Company, the Drugs, and the Inventors Today 67

4 Discovery of Lipitor 71
 Splendid History 71
  Parke, Davis & Company 71
  Dilantin 74
  Chloromycetin 76
  The pill 78
  Lopid 80
 Roger S. Newton 81
 Bruce D. Roth 84
 The Discovery of Lipitor 86
  Shoulders of giants 86
  The first success... and disappointment 90
  The second refrain 92
  CI-971 94
  CI-981 96

5 Development of Lipitor 101
 Checkered Life 101
 Chiral or Not, You Have to Make It First 105
 Clinical Trials—Divine Providence? 108
 Canter and the FDA 113
 What's in a Name? 120
 A Marketing Partner 121

6 To Market, to Market 125
 Pfizer Inc. 126
  The small drug firm in Brooklyn made big 126
  Penicillin 127
  Drugs of its own 128
 Windfall: The Era of Blockbusters 131
  Viagra 132
  Zoloft 134
  Diflucan 135
  Zithromax 137
  Norvasc 138

    The Merger   140
    PROVE-IT   142
    The Patent Litigation   145

7 Baycol, Crestor, and Cholesterol Drugs beyond Statins   149
    The Story of Baycol   150
    Crestor   152
    Cholesterol Drugs beyond Statins   155
        Zetia and Vytorin   155
        Torcetrapib, a bitter disappointment   158
        ApoA-1 Milano and Newton   162

8 Reflections   167
    Triumph of the Heart   167
        Statins set a high standard in efficacy   168
        Statins set a high standard in safety   169
        Statins set a high standard in financial benefits   169
    The Drug Industry and the Blockbuster Model   170

Appendix: Drug Names   173

Notes and References   175

Index   189

TRIUMPH OF THE HEART

CHAPTER I

Cholesterol

The story of statins starts with cholesterol because statins are a class of drugs that reduce low-density lipoprotein (LDL) cholesterol, the "bad" cholesterol. LDL cholesterol, in turn, is a major risk factor for coronary heart disease, the leading cause of death worldwide and projected to remain so through 2025. About 1.5 million Americans suffer heart attacks each year, and heart disease has emerged as the biggest cause of death in the United States, killing 911,000 people in 2003.

Before the 1940s, the average lifespan in America was 47 years, and heart disease did not contribute to mortality to a large extent because people often died of infections. Currently, an average American lives to celebrate her 77th birthday. As a consequence, heart-related disease has risen to be the number one killer. Coronary heart disease manifests in many forms: angina, arrhythmia, atrial fibrillation, congestive heart failure, hypertension, atherosclerosis, myocardial infarction (heart attack), and sudden cardiac death. Atherosclerosis, or blockage in arteries, results when a buildup of cholesterol, inflammatory cells, and fibrous tissue called plaques forms on an artery wall. If these plaques rupture, they can block blood flow to critical organs such as the heart or brain and can lead to heart attack or stroke.

Despite the many different forms of cardiovascular disease, the molecule cholesterol is a common denominator for most of them. Therefore, in order to understand coronary heart disease, we first need to take a look at the cholesterol molecule.

The Janus-Faced Molecule

According to Roman mythology, Janus is the guardian of portals and patron of beginnings and endings. Just like the two-faced Roman god, cholesterol is a double-edged sword for the human body. On the one hand, it is an essential building block for many crucial ingredients the body needs. On the other hand, it can be lethal when it forms plaques on the surface of the arteries and subsequently causes coronary heart disease.

Make no mistake, cholesterol is vital to our existence. It is most abundant in our brains—23% of total body cholesterol resides there, making up 1/10th of the solid substance of the brain. Red blood cell membranes are also rich in cholesterol, which helps stabilize the membranes and protect the cells. In addition, cholesterol is necessary for producing bile acids that help digest fats in our food. Moreover, it is the precursor molecule for the synthesis of sex hormones, including progesterone, testosterone, and estrogen. In fact, *all* steroids in the body are derived from cholesterol, which is converted into specific hormones by biochemical transformations catalyzed by enzymes. For instance, enzymes in testes convert cholesterol to testosterone, whereas enzymes in ovaries convert cholesterol to estrogen. Other enzymes in the body convert cholesterol to additional hormones, such as cortisol, a hormone secreted by the outer layer of the adrenal glands when we are under stress. Addison's disease is due to adrenal cortex deficiency and is characterized by the failure of the adrenal glands and the inability to produce cortisol.

It is apparent that cholesterol is an indispensable ingredient for life. But like everything else in life, too much of a good thing can be bad. By now, the experimental, genetic, and epidemiologic evidence all point to elevated cholesterol levels as a major risk factor for cardiovascular disease.

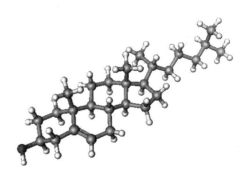

Figure 1.1 Molecular structure of cholesterol.

Cholesterol helps plaque build up, which constricts or blocks arteries and leads to angina, heart attack, and stroke.

Cholesterol in the human body comes from two sources. One is from intestinal absorption of dietary cholesterol. The other is generated inside the body, primarily in the liver, to meet the body's need if the diet is lacking sufficient quantities. The liver makes about 70% of the body's cholesterol—our bodies produce, on average, three or four times more cholesterol than we get from dietary sources.

Since it was first isolated from gallstones in 1784, cholesterol has fascinated researchers from many areas of science and medicine. Thirteen Nobel Prizes have been awarded to scientists who devoted major parts of their careers to cholesterol research.

Believe it or not, cholesterol comprises about three-quarters of a gallstone—the remainder is calcium salts. French physician-chemist François Poulletier was the first to obtain pure cholesterol from gallstones. At first, Poulletier erroneously identified cholesterol as a form of wax. Some 30 years later, French chemist Michel E. Chevreul shattered that notion by showing that cholesterol could not be saponified—wax is composed of esters and thus can be saponified. Chevreul named it cholesterine ("solid bile," from the Greek *chole* for bile and *stereos* for solid).

The exact empirical formula of cholesterol was accurately established in 1888 by Austrian botanist Friedrich Reinitzer,[1] who worked at the Imperial Institute for Plant Physiology at the German University in Prague. Interested in the biologic roles of cholesterol in plants, Reinitzer initially studied cholesterol isolated from the carrot root. However, its cholesterol content was so minute that Reinitzer resorted to purchasing cholesterol from a factory. After purifying the sample by treatment with alcoholic sodium hydroxide, Reinitzer treated cholesterol with bromine and obtained a compound that "precipitates out as *splendid crystals*."[1] Using a rudimentary but reliable method called elemental analysis involving combustion of the compound and then analysis of the carbon and hydrogen contents, he deduced the precise molecular formula. In his publication in the prestigious German chemistry journal *Monatshefte für Chemie* (*Chemical Monthly*) in 1888, Reinitzer was very confident: "The formula of cholesterol *must read* $C_{27}H_{46}O$."[2] Chevreul was still alive at the time of Reinitzer's publication and lived another year until he died at the age of 103.

The cholesterol molecule has four rings, which made its structural elucidation an extreme daunting challenge, occupying scientists for a good part

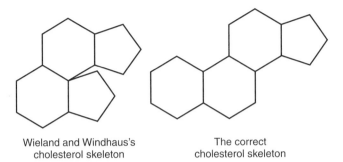

Wieland and Windhaus's cholesterol skeleton

The correct cholesterol skeleton

Figure 1.2 Structure of cholesterol.

of the first quarter of the twentieth century. Proof of cholesterol's structure was obtained chiefly through the brilliant work of two Germans, Heinrich O. Wieland in Munich and Adolf Windaus in Göttingen. They received the Nobel Prize in Chemistry in 1927 and 1928, respectively, for their work on cholesterol.[3] As an indication of the intricacy of cholesterol's structure, the initial structure suggested by Windaus and Wieland in their Nobel lectures was actually incorrect![4] But this does not detract in any way from their contributions, because at the time, few modern instruments were at their disposal. Nuclear magnetic resonance (NMR) spectroscopy, without which a modern organic chemist would easily be lost, was not invented until the 1940s, commercialized in the 1950s, and not widely available until the 1960s.

Adolf Windaus[3] was the son of a drapery manufacturer in Berlin. In his youth, he apprenticed under Emil Fischer, the 1902 Nobel laureate in chemistry. Like many great scientists, Windaus exhibited limitless curiosity and energy. He pursued zoology in addition to his medicinal and chemical interests. Acting on a suggestion by the well-known chemist Heindrich Kiliani, Windaus began working on cholesterol and related compounds in 1901. He also initiated collaboration with his friend Heinrich Wieland, who had already isolated cholanic acid, a cholesterol analog and synthetic precursor of bile acid. In addition, Windaus reported in 1910 that atherosclerotic plaques contained six times more free cholesterol than a normal arterial wall and 20 times more esterified cholesterol. His observations were among the first published linkages between cholesterol and heart disease.

Figure 1.3 Steroid structure on the commemorative stamp for Barton's Nobel Prize © Royal Mail.

Windaus won the Nobel Prize in Chemistry in 1928 for services rendered through his research into the structure of the sterols and their connection with the vitamins. A German with a conscience, Windaus was opposed to Nazi ideology. As head of the Institute for Organic Chemistry at Göttingen, he protected a Jewish student from dismissal. In the library of the University of Göttingen, a copy of Adolf Hitler's *Mein Kampf* had handwritten in it: "by order of the government." Many believed that the handwriting was that of Windaus. His profound moralistic conscience compelled him to refuse to carry out research on poison gas as far back as World War I. Windaus voluntarily stopped all scientific research in 1938 at age 62 for his refusal to concede to the National Socialists. Windaus's friend Wieland was not a big fan of the Nazis, either. He isolated pigments from 200,000 butterflies, elucidated their novel structures, and named them pteridines. In a conference with numerous brown-shirted students in the audience he remarked: "I will be unable to continue this research since the Government regards the collecting of butterflies as cruelty to animals, incompatible with the ethics of the National Socialist Party. I can only hope that this research will be carried out by my colleagues abroad."[3]

Figure 1.4 Adolf Windaus © Nobel Foundation.

Figure 1.5 Heinrich Wieland © Nobel Foundation.

By 1932, Wieland finally proposed the correct structure of cholesterol. Its true structure was unambiguously established in 1945, based on X-ray diffraction data, by Dorothy Crowfoot Hodgkin at Oxford University. Hodgkin was awarded the 1964 Nobel Prize in Chemistry for her tremendous contributions to elucidating structures of complicated molecules such as cholesterol and insulin.

During World War II, cholesterol, the starting material for the preparation of steroid hormones, was in short supply. The Allies resorted to extracting

Figure 1.6 Dorothy Hodgkin © Royal Mail.

CHOLESTEROL 7

solanine from potato sprouts and then converting solanine to cholesterol. If cholesterol could have been synthesized from less complicated building blocks, it would have been a great boon to the Allies. In fact, deducing the structure of cholesterol was difficult, but the synthesis would prove to be even more challenging. A fierce competition to finish the synthesis first raged between two prominent organic chemists of that time: Sir Robert Robinson of Oxford University and Robert Burns Woodward of Harvard University. As luck would have it, Robinson and Woodward finished the first total syntheses of cholesterol almost simultaneously in 1951. It was a crown jewel of twentieth-century synthetic chemistry. Robinson, Woodward, and John W. Cornforth all went on to receive Nobel Prizes in Chemistry in 1947, 1965, and 1975, respectively. Cornforth was Robinson's student who took over the cholesterol project and finished its total synthesis.

Robert Robinson was born in Chesterfield, England, the son of a highly successful manufacturer of surgical dressings.[5] Robinson went to the University of Manchester to study chemistry under William Perkin, Jr., whose father, William Perkin, Sr., was the first organic chemist to become very wealthy by inventing, manufacturing, and selling mauveine, a purple dye. After receiving his Ph.D., Robinson took a junior post at the University of Manchester, where he was strongly influenced by Arthur Lapworth, from whom he developed an interest in the theoretical aspects of organic chemistry. From 1912 to 1916, Robinson taught at the University of Sydney, after which he returned to the University of Liverpool, where he developed the famous "Robinson annulation reaction" now commonly taught in sophomore organic chemistry. In 1930, Robinson moved to Oxford University, where he was appointed head of the Dyson-Perrin Laboratory. He turned the laboratory into a center for excellence in organic chemistry, and students from all over the world flocked to Oxford to study with him.

By 1932, just as the structure of cholesterol had been correctly elucidated, Robinson decided to embark on his quest for the synthesis of steroids, a class of compounds including cholesterol. In 1939, he assigned the cholesterol synthesis to Australian graduate students John W. "Kappa" Cornforth and Rita Cornforth. John Cornforth began losing his hearing when he was 10 and was totally deaf by his mid-20s. His wife and fellow student, Rita, was the major channel of communication between John and his colleagues. After graduating from Sydney University, the Cornforths went to Oxford to pursue their Ph.D. degrees in organic chemistry, since none was offered in Australia at the time. John Cornforth worked on cholesterol through his Ph.D. studies until 1941. Because Howard Florey and Ernst Chain at Oxford were working on the miracle drug penicillin, Robinson and Cornforth halted their cholesterol efforts and joined the fray of penicillin synthesis, although without much success. After the war, Cornforth could not take a teaching post due to his deafness, so he moved to the National Institute for Medical Research in London. While there, Robinson arranged with the director, Charles Harrington, for John Cornforth to continue working on the cholesterol project.

Despite his brilliance, Robinson was a micromanager, with a tendency to dismiss ideas from his collaborators if they were not in line with his own thoughts. He liked to run a tight ship and was always reluctant to delegate responsibilities to even the most qualified colleagues. Therefore, a student with independent ideas had no choice but to leave Oxford and start on his own, leading to a "brain drain" in the Dyson-Perrin Laboratory. In 1950, Cornforth persuaded Robinson to let him take over the cholesterol project in the interest of speed, and Cornforth finally completed the project one year later. Interestingly, across the Atlantic Ocean, Professor Woodward at Harvard University finished his version of the total synthesis of cholesterol at virtually the same time.

Two giants of organic chemistry working on the total synthesis of the same molecule undoubtedly created a certain amount of competition and tension. The apocryphal showdown between Robinson and Woodward took place in 1951. According to Sir Derek Barton, who was a friend to both of them:

> By pure chance, the two great men met early on a Monday morning on an Oxford train station platform in 1951. Robinson politely asked Woodward what kind of research he was doing these days; Woodward replied that he thought that Robinson would be interested in his recent total synthesis of cholesterol. Robinson, incensed and shouting "Why do you always steal my research topic?" hit Woodward with his umbrella. This story must be true for Woodward told me about it several days later. Fortunately, after that incident, the two great men became firm friends, each recognizing the brilliance of the other. Robinson finally admitted that he had met someone who was his equal.[6]

Figure 1.7 Robert Robinson © Nobel Foundation.

Figure 1.8 Derek Barton © Nobel Foundation.

Robert Burns Woodward was born in Boston, Massachusetts.[7] His mother, Margaret Burns, a Glasgow native, named her only son after the Scottish poet Robert Burns. His father, Arthur Woodward, of English descent, died in the great influenza epidemic of 1918 when Robert Burns (later known as "R. B." to his friends) was only two years old. In high school, Woodward performed almost all of the experiments in Ludwig Gattermann's *Practical Methods of Organic Chemistry*. He went to MIT in 1933 at age 16, and two years later he had taught himself more chemistry than the professors at MIT had acquired in their lifetimes. Fortunately, the professors at MIT eventually recognized Woodward's genius and allowed him to take examinations without attending their classes. In his junior year at MIT, he signed up for 186 credit hours of classes. Since there are only 168 hours in a week, even counting nights and weekends, he must have skipped a few classes. In reality, Woodward simply took the exams without going to the classes and completed his B.S. and Ph.D. in only four years. After graduation, he taught summer school at the University of Illinois, where his arrogance alienated and outraged some of the most prominent organic chemists of the time, including the legendary Roger Adams (best known for the eponymous Adams' catalyst and the founder of *Organic Syntheses*). In the fall of 1937, Woodward moved back to Harvard as a research assistant to Professor E. P. Kohler. Apparently a genius with a prodigious memory, he rapidly rose to full professor in 1950.

Woodward was larger than life. He was a heavy drinker and a chain smoker and always lived life on the edge. His peculiarity was not limited to his omnipresent blue suit and blue tie. His beloved Mercedes sedan was, of course, in his ubiquitous blue, and his students even painted his parking space blue. He was well known for his colored chalk talks where he enjoyed starting at the upper left-hand corner of a very large blackboard and finishing at the lower right-hand corner.

As a testament to his skills as a synthetic chemist, Woodward completed his total synthesis of cholesterol in just two years. Two main factors contributed to the completion of such a daunting task in a short time.[8] One was Woodward's genius in conceiving an elegant synthetic plan. The other factor was his ability to enlist help from the drug industry. Not only did Merck and Monsanto provide funding for Woodward's postdoctoral fellows who worked on the cholesterol synthesis, but Monsanto also produced an advanced intermediate tricyclic ketone on a pilot plant scale, furnishing Woodward with a virtually unlimited supply.

Because of Woodward's high-pressure style of leadership, the four postdoctoral fellows on the cholesterol project worked night and day, weekends and holidays. As Woodward's noon-to-3 A.M. hours became commonplace, his eager disciples followed suit in increasing droves, so midnight was high noon for the Woodward group. His weekly group meeting invariably went on until long after midnight. On Christmas day, 1950, they successfully synthesized a key intermediate ketone. An exuberant Woodward promptly christened it "Christmasterone."[8]

Right after Woodward's completion of his cholesterol total synthesis, Merck announced its intention to synthesize cortisone, a drug for rheumatoid arthritis, using Woodward's chemistry. The optimism of Merck, in turn, led to the heralding of the Woodward synthesis on American radio: "It now seems reasonable to expect mass production of this wonder-working hormone within a short time, as a result of the greatest international race in modern chemistry—the race to achieve the total synthesis of a steroid."[8]

Unfortunately, American media was overly optimistic in that account. In the end, Woodward's complicated total synthesis of cholesterol did not contribute to the treatment of even one arthritic patient.[9] Instead, scientists Durey H. Peterson and H. C. Murray from the Upjohn Company discovered a highly innovative approach using microbiologic fermentation technology in steroid chemistry, which enabled a commercially viable route for synthesis of cortisone in 1949.

One of the 10 Greatest Discoveries in Medicine

The discovery of the linkage between cholesterol and atherosclerosis is widely viewed as one of the 10 greatest discoveries in medicine.

Nikolai N. Anitschkov, a Russian Army physician, is credited as the first to make the connection between cholesterol and atherosclerosis.

Anitschkov, a Russian aristocrat, was born in 1885 in St. Petersburg.[10] He earned his doctorate in 1912 from the Imperial Military Academy of Medicine in St. Petersburg. In 1908, in the same medical school, a clinician named A. I. Ignatowski had induced atherosclerosis in rabbits. After feeding rabbits only meat (rabbits are herbivores), eggs, and milk, Ignatowski observed plaques in rabbit aortas in just a few weeks. He erroneously concluded that the plaques were caused by proteins. He published a paper explaining that protein was toxic to young rabbits and let the matter rest.

But Anitschkov was not satisfied with Ignatowski's simplistic account. He dug deeper and tried to look at the molecular level. Anitschkov and his student S. Chalatov fed rabbits purified cholesterol dissolved in sunflower oil through a stomach tube for 3–4 months. In due course, they observed cholesterol-induced vascular lesions closely resembling those of human atherosclerosis, both grossly and microscopically. In contrast, the control group of rabbits fed only sunflower oil did not show lesions. In 1913, Anitschkov and Chalatov published their legendary article in the prestigious German journal *Zentralblatt für allgemeine Pathologie und pathologische Anatomie* with the title "On Experimental Cholesterin Seatossi and Its Significance in the Origin of Some Pathological Processes." That year, Anitschkov had just turned 28 and was fresh out of medical school. Later, he reproduced the same results in guinea pigs, although he was not successful in inducing atherosclerosis in rats and dogs. He summarized his discovery succinctly: "Without cholesterol there can be no atherosclerosis."[11]

On a personal level, Anitschkov embodied the refined characteristics of a cultured gentleman of the Old Russian era. In 1917, after the October Revolution, Anitschkov joined the Bolshevik party and remained a staunch communist for the rest of his life. He was a good friend of Soviet leader Joseph Stalin and was promoted to lieutenant general, the highest rank obtainable in the Soviet Army Medical Corps. Regrettably, when he died in 1964, only a handful of his contemporaries took notice of his death and his contributions.

The Western medical society at large, having its own preconceived ideas, did not heed Anitschkov's progressive conclusions for several reasons. First, scientists and physicians were not convinced that Anitschkov's rabbit data would translate to humans. Rabbits are herbivores, and humans are omnivores. The data from dogs would be more relevant, but Anitschkov himself did not observe atherosclerotic plaque formation in dogs. It turns out that one has to block thyroid function in order to observe lesions caused by cholesterol in dogs. To date, atherosclerosis has been induced in many species: rabbit, guinea pig, pigeon, chicken, dog, pig, and monkey. Interestingly, scientists have still not been able to induce atherosclerosis in rats by feeding them cholesterol. As recently as 1999, a group of prominent pathologists in atherosclerosis research still argued that Anitschkov's rabbit model is not identical to human atherosclerosis.

A less profound reason for Anitschkov's lack of recognition could be as simple as the language barrier. In the English language alone, Anitschkov's last name has been spelled in numerous ways: Anitchkov, Anitchkow, Anichkov, and Anichhkow, to list just a few.[12] This has created tremendous confusion in the literature. By human nature, one is often reluctant to recognize anybody or anything if one does not know how to pronounce or spell the name.

Another underlying reason that Anitschkov's results were largely ignored could be that coronary heart disease was a less urgent issue compared with infectious diseases in the first half of the twentieth century. Anitschkov discovered the link between cholesterol and atherosclerosis 40 years ahead of the rest of the medical world using simple experiments and marshaling his power of deductive reasoning. Looking back at Anitschkov's scientific achievements, today we can easily recognize his genius. Considering our present-day preoccupation with prevention and treatment of heart disease, it is easy to see why Anitschkov's keen observation that cholesterol caused atherosclerosis in rabbits is now considered one of the 10 greatest discoveries in medicine.

The Good, the Bad, and the Ugly

So, how does cholesterol accumulation cause atherosclerosis? Since cholesterol is greasy and does not dissolve in the blood, it latches onto water-soluble "carrier" proteins, forming lipoproteins. As the name implies, a lipoprotein is a molecule consisting of a lipid (in this case, cholesterol) and a protein. Because of different possible protein carriers, not all cholesterols are created equal. In fact, they can have opposing effects on the heart. Cholesterol in low-density lipoprotein (LDL), often known as the "bad" cholesterol, is the fundamental source of blood cholesterol to body cells. It can slowly build up in the walls of the arteries feeding the brain and heart and can form plaques. The average American has an LDL cholesterol level of 127 mg/dL, whereas the American Heart Association recommends that it be lower than 100 mg/dL. On the other hand, cholesterol in high-density lipoprotein (HDL), frequently dubbed the "good" cholesterol, is the form by which cholesterol is taken away from the arteries and brought to the liver, where it can be removed from circulation. The higher the HDL-cholesterol levels, the better for your heart. In general, women have higher levels of HDL, which may explain why women have longer life expectancies than men. The presence of estrogen is somehow related to higher HDL-cholesterol levels. After menopause, when estrogen secretion ceases, the relative rates of heart disease in women are the same as in men.

Historically, physicians and chemists have made the most contributions to medicine. However, a few physicists have also revolutionized medical science from time to time. One of the most prominent examples is Wilhelm C. Röntgen's discovery of X-rays. Another case is the discovery by biophysicist John W. Gofman, at the University of California at Berkeley, that lipoproteins with different densities have different effects on the heart.

By now, the terms LDL and HDL cholesterol have become part of our daily vocabulary. Gofman first discovered that cholesterols in LDL and HDL have different effects on the heart. Born in Cleveland, Ohio, in 1918, Gofman was the son of Russian Jewish immigrants.[13] He studied chemistry at UC Berkeley in 1940, where he took part in the Manhattan Project. After earning his Ph.D. in physics under the direction of Nobel laureate Glenn T. Seaborg, he entered medical school at University of California at San Francisco, earning his M.D. degree in 1946. After completing his internship, Gofman joined the Berkeley faculty as an assistant professor of medical physics in 1947.

As a young assistant professor at Berkeley, Gofman was supposed to carry out some independent research on cancer. He pondered this task for more than six months but did not come up with any good ideas. During that time, though, there was a lot of controversy as to whether blood cholesterol had any impact on heart disease. Unlike his peers, Gofman took Anitschkov's work very seriously. Along with his colleague Frank T. Lindgren, he decided to study how blood transported cholesterol. As luck would have it, analytical ultracentrifuges had just become available around that time. Gofman purchased one of only two available instruments despite its astronomical price tag of $16,000. The powerful centrifuge could spin as fast as 80,000 revolutions per minute (rpm) and generate forces that were 300,000 times that of gravity. Using this machine, lipoproteins could be separated according to their density. In 1948, Gofman and his students began to study rabbit blood serum but could not get consistent protein patterns, even though the serum had very high cholesterol levels. Then they switched to the investigation of lipoproteins, but these also initially proved problematic. However, after weeks of probing, Gofman and Lindgren discovered that the solution was very straightforward—simply adding a little salt would make the lipoproteins float to the top after ultracentrifugation. From 1949 onward, their flurry of experimental results solidly established that lipoproteins were predictive of heart disease. While LDL cholesterol causes atherosclerosis, HDL cholesterol actually protects the heart.

During the Korean War in 1951, the Pentagon dispatched pathologists to Korea to autopsy 2,000 young, healthy U.S. soldiers killed in action. They performed detailed dissections of about 300 corpses and found that more than three-quarters of all the young men with an average age of 22 had evidence of lesions in their arteries. They had discovered unexpected evidence of diseased coronary arteries. Two years later, Major William Enos and Lieutenant Colonel Robert H. Holmes described the autopsy evidence in the *Journal of the American Medical Association* (*JAMA*), which galvanized Americans' awareness of heart disease.

In 1956, University of Minnesota physiologist Ancel Keys, the inventor of the K-ration ("K" for Keys) during World War II, proposed that a high-cholesterol diet was a major risk factor for coronary artery disease.[14] But his idea received more skepticism than support from his peers. In order to refute early doubts about the links between heart disease, cholesterol, and diet, Keys undertook an immense international collaboration involving seven countries: Italy, Greece, Yugoslavia, the Netherlands, Finland, the United States, and Japan. The study began in 1958 and involved 12,000 healthy middle-age men. The "Seven Countries Study" demonstrated soundly that the types of fat eaten as food affected cholesterol in the blood and predicted death from heart attacks and strokes. While Japan and Italy had the lowest number of cardiac events, the United States and Finland had the highest. The respective diets obviously played an important role. To illustrate how rich in fat the Finnish diet was, they were known to butter their cheese! As a consequence, the Finnish were 14 times more likely to be inflicted by coronary heart disease than the Japanese.

In 1959, Keys and his wife, Margaret, wrote a cookbook, *Eat Well and Stay Well*, promoting the Mediterranean diet. The book turned into an unexpected bestseller. Using the royalties from the book, they bought a house in Naples, Italy, where coronary disease was "nonexistent" at the time. The Seven Countries Study earned Keys the nickname "Mr. Cholesterol" and a cover story in *Time* magazine in January 1961. Keys lived to be 100 years old, thanks at least in part to the healthy diet that he himself preached and practiced.

Hitler's Gift

With the first laboratory total syntheses of cholesterol achieved by Robinson and Woodward, one of the most brilliant chapters of organic chemistry came to a close. Needless to say, their designs did not and could not imitate Mother Nature's divine design for accomplishing the same feat.

Now, how exactly is cholesterol synthesized in the human body? Finding the answer took scientists around the world several decades, earning them several Nobel Prizes along the way. A large body of knowledge was accumulated in the 1950s and 1960s, chiefly from the laboratories of Konrad E. Bloch at Harvard, George Popják at UCLA, John W. Cornforth at Oxford, and Feodor Lynen at the Max Planck Institute in Germany. We now know that cholesterol is synthesized in the body from acetic acid, a molecule with only two carbons and the principal ingredient in the vinegar that we use every day in the kitchen.

Konrad Bloch was Hitler's gift to American science.[15] Born in Neisse, Germany (now Nysa, Poland), Bloch attended Munich University for his undergraduate education. At Munich, he was under the great influence of Hans Fischer and Richard M. Willstätter, both Nobel laureates in chemistry.[16] In 1934, Bloch obtained a master's degree in chemical engineering. The brutal Nazification of Germany hindered Bloch's pursuit of further education because he was Jewish. The authorities told him that he was denied entry into graduate school because Fischer did not want to take him as a student. Of course, that was a total lie. Fischer was actually sympathetic and arranged for Bloch to study in Switzerland. During Bloch's stay in Switzerland, he went to the Eidgenössische Technische Hochschule (Federal Institute of Technology) in Zürich, where he approached Leopold Ruzicka, who would win the Nobel Prize in 1939. However, Ruzicka did not think too highly of Bloch and blatantly told him that he did not have the qualifications to earn a Ph.D. in organic chemistry. Disappointed but not daunted, Bloch went on to learn biochemistry in Davos, Switzerland. While studying the tubercle bacillus, the bacterium that causes tuberculosis, Bloch found no cholesterol in it, which contradicted the findings of Rudolph J. Anderson at Yale University. Educated in the German system, where professors were revered like gods, Bloch gathered all his courage and wrote a letter to Anderson, a foremost authority on lipids of this organism. To his surprise, Anderson was extremely gracious, writing to him and admitting that Bloch's preparation was probably purer, thus more reliable. Anderson also helped Bloch get a visa to the United States, probably saving his life because his permission to stay in Switzerland would soon expire.

Arriving in America in 1936, Bloch took Anderson's advice and applied to graduate school at Columbia University, where he was offered a position to work under Hans T. Clarke. One reason that Clarke took him was probably because Bloch played cello and Clarke was a big fan of chamber music. Laudatory letters from Fischer and Willstätter did not hurt his chances, either. With Clarke, Bloch completed a relatively straightforward piece of research on amino acid chemistry and obtained his Ph.D. within a year and a half. In 1938, Bloch applied for a postdoctoral position under the gifted Rudolf Schoenheimer, also a Jewish refugee from Germany, in the same department at Columbia. Schoenheimer was initially hesitant to hire Bloch because his Ph.D. thesis was "thin." Apparently, Schoenheimer's reluctance quickly dissipated, though, because he hired Bloch, who proved himself to be very productive, publishing several papers in a short period of time.

Under Schoenheimer, Bloch also began to study the biosynthesis of cholesterol using a new method called the isotope tracer technique, which uses compounds labeled with deuterium in place of hydrogen. In 1941, Schoenheimer committed suicide because of his depression, and his work had to be divided among three postdoctoral fellows by drawing lots. Bloch ended up with the cholesterol problem by the luck of the draw, literally.

Previously, two German researchers, R. Sonderhoff and H. Thomas, had obtained preliminary results implicating acetic acid as an important precursor to cholesterol. Using deuterium-labeled acetate, Bloch was able to show how specific parts of cholesterol are formed from the specific atoms of acetic acid.

At Columbia, Bloch carried out an experiment to demonstrate that cholesterol was a precursor to the steroid hormones. He wanted to take advantage of the relatively massive urinary excretion of pregnanediol, a progesterone metabolite found in the late stages of human pregnancy. This was a promising, if somewhat unorthodox, approach. He attempted to persuade the Department of Obstetrics and Gynecology to provide him with suitable patients, but his request was brusquely denied. Since his wife Clore was pregnant, Bloch gave her deuterium-labeled cholesterol and analyzed the steroid hormone content of her urine samples. Here is an excerpt of his paper published in 1945:

> Deutero-cholesterol was ingested by a woman in the eighth month of pregnancy, the progesterone metabolite isolated from the urine excreted during the experimental period in the form of pregnanediol glucuronidate, and analyzed for deuterium. Significant concentrations were present in the samples showing clearly that pregnanediol had been formed directly from cholesterol.[17]

One month later, his son Peter was born, identifying the essential collaborator in this experiment.

A year later, Bloch moved to the University of Chicago to start his own independent research group. In 1954, Harvard decided to grow its Biological Sciences Department and hired a number of outstanding professors away from Chicago, including Bloch, Frank Westheimer, and Eugene Kennedy.

Another great leap forward in deciphering the cholesterol biosynthesis pathway was identification of squalene as a precursor to cholesterol by Robert Langdon, a Ph.D. student of Bloch's at Chicago in 1955. Meanwhile, Feodor Lynen of Germany, Adolf Windaus's son-in-law, clarified many of the steps leading from acetate to squalene. Two years later, Bloch and his associate T. T. Tchen proved that molecular oxygen, the oxygen present in the air, was the source for the only oxygen atom in the cholesterol molecule—the hydroxyl group.[17]

After a conversation with R. B. Woodward, Bloch proposed a possible pathway from squalene to cholesterol. He carefully evaluated how squalene, a linear molecule, folded and modified to produce cholesterol, a four-ring structure, because the transformation was not obvious at first glance. With great insight, Bloch proposed that lanosterol, a sterol found in wool fat, might be an intermediate. More important, he carried out tracer studies supporting this hypothesis.[18]

By 1959, Bloch and his coworkers decisively established the "acetic acid → squalene → cholesterol" cascade for cholesterol biosynthesis. They determined that the transformation from acetic acid to cholesterol is a complicated sequence involving 36 steps and just as many enzymes to facilitate the process. Bloch shared the 1965 Nobel Prize in Physiology or Medicine with Lynen for discoveries concerning the biosynthesis of cholesterol by the body from acetic acid. By the time of his retirement in 1982, Bloch's laboratory at Harvard had become the Mecca of lipid research.

During the Vietnam War, Bloch signed a petition protesting chemical warfare in Vietnam. He said, "The fact that you are a Nobel laureate should be irrelevant. But it never is. When these letters are gotten together, they choose people who sound impressive to the public."[18] Even after his retirement, Bloch never ceased to do research, although not in the laboratory. In 1994, he published a book titled *Blondes in Venetian Paintings, the Nine-Banded Armadillo, and Other Essays in Biochemistry*. In his dedication, Bloch wrote, "To the late Rudolph J. Anderson, professor of biochemistry at Yale University, who facilitated my immigration to the United States."[19] Interestingly, the book was published by Yale University Press rather than Harvard University Press, which felt the collection of essays was too technical for a lay audience.

The Framingham Heart Study

In the wake of President Franklin D. Roosevelt's death from a massive stroke on April 12, 1945, the U.S. Congress allotted half a million dollars to the U.S. Public Health Service for a study that would investigate the causes of heart disease. Framingham, Massachusetts, a small town 18 miles west of Boston, was chosen for the long-term epidemiologic study. One reason it was chosen was that 5,000 Framingham residents had already taken part in the Framingham Tuberculosis Demonstration Study from 1916 to 1923. Another reason was that Harvard Medical School was becoming a leader in cardiology, and some key players, such as Harvard cardiologist David Rustein, wished to have a study site close by. The study enrolled Framingham's 5,209 residents between 30 and 60 years of age. Every two years, the volunteers received a careful medical examination and a battery of tests. The systemic epidemiology studies included routine health monitoring and blood sampling, measuring the incidence of heart attacks and stroke, and determining various associated risk factors. Initially, the U.S. Public Health Service appointed Gilcin Meadors as the director of the study. Meadors, a young southern bureaucrat, laid a solid foundation for the study but did not get along with the staff and scientists. In 1950, the study changed hands from the U.S. Public Health Service to the National Heart Institute, a division of the National Institutes of Health (NIH). Thomas "Roy" Dawber, a practicing physician and native New Englander, replaced Meadors as director of the study, a job he would hold for almost 20 years.

Soon after John Gofman at the University of California at Berkeley was able to separate lipoproteins as LDL and HDL using an analytical ultracentrifuge in 1949, the Framingham researchers began sending samples to Gofman for lipid density measurements. The study decisively confirmed that escalating levels of LDL cholesterol are a key risk factor for plaque formation in artery walls, and HDL cholesterol is the principal protective component.[20] In 1977, the Framingham study also determined that increased levels of triglycerides and LDL increase the risk of coronary heart disease, although cholesterol level elevation seemed to carry a greater risk than triglyceride level elevation does. Triglyceride, like cholesterol, is also a lipid.

In April 1957, Dawber and his colleague William B. Kannel published their landmark article "Coronary Heart Disease in the Framingham Study" in the *American Journal of Public Health*. Their results showed that escalating cholesterol levels were proportionally correlated to increased risk for heart disease. They found that each 1% increase in cholesterol levels leads to a 2% increase in the incidence of coronary heart disease. Dawber and Kannel were also the first to use the term "risk factor" in the medical literature in 1961. Since then, the Framingham study has identified many major risk factors, including smoking (1962), sedentary lifestyle (1967), hypertension (1970), and diabetes (1974), in the development of coronary heart disease.[21]

In 1968, the Framingham Heart Study reached its initial tenure of 20 years. Despite its illustrious achievements, the epidemiologic study lacked "glamour and flamboyance,"[20] and many believed that it had accomplished much of what it had set out to do. The director of the NIH decided to close it down in 1969, against the recommendation of the review committee. Many renowned cardiologists, including Paul Dudley White of Massachusetts General Hospital, protested. White even wrote a letter to President Richard Nixon and asked him to intervene. Dawber, fearing the worst, began raising private funding for the study, including from the insurance industry. The Framingham Heart Study survived and became a collaboration between government and academia.

Since 1971, Boston University School of Medicine has been the recipient of a contract with the National Heart, Lung, and Blood Institute to provide examinations and extensive analytical and scientific support for the Framingham study.[20] That year, the study recruited 5,124 children of members of the original study group and their spouses for the "Offspring Study." The ensuing two decades were an era when deaths from heart attacks and strokes fell sharply while the costs of cardiovascular care continued to soar. The Framingham offspring, now middle-age, are in many respects significantly healthier than their parents were when they entered the federally financed study 44 years ago. Their blood pressure and cholesterol levels are considerably lower, and far fewer participants smoked cigarettes. But the men are heavier and less active, and both the men and the women have much higher rates of diabetes than their parents did at comparable ages. "Since the early 1970s, when the offspring study began, there has been nearly a 40 percent drop in the death rate from heart attacks and a 58 percent decline in the death rate from strokes in the United States," observed Dr. William Castelli, director of the Framingham Heart Study, in 1994, "but the rate at which people suffer heart attacks and strokes has not fallen to a comparable degree."[22]

Dr. Kannel, who had been monitoring changes among the Framingham offspring, said, "Currently, twenty percent of the 50-year-olds, forty percent of the 60-year-olds and fifty percent of the 70-year-olds in Framingham are taking drugs to lower their blood pressure. When you get to numbers like these, you might as well put the drugs in the reservoir."[22]

In 2002, the "Third Generation Study" kicked off. Six decades and more than 1,300 scientific papers later, the risk factors behind cardiovascular disease are common knowledge to most of us, thanks to this study.[23]

In early 2005, Framingham Heart Study researchers reported interesting data supporting the validity of the old adage that what's bad for your ticker may be good for your bean. A team of scientists at Boston University found an association between *naturally* high levels of blood cholesterol and better mental function. The results are not completely surprising, considering the brain has the highest concentration of cholesterol (23% of the brain) compared to other body parts. Furthermore, cholesterol is important for brain development in infants and plays a role in how neurons function in adults. Nonetheless, if one assumes that the higher the cholesterol levels the better, one would be wrong. Just like everything else in life, a balance needs to be reached to enjoy both good physical and mental health.

The Framingham Heart Study has now followed three generations of participants. Framingham is now known as the town that changed America's heart, and the Framingham Heart Study is considered the crown jewel of the world's foremost medical research institution. Following this study, medical research began to focus on how to reduce cholesterol levels.

The Receptor

In the pantheon of scientists, many have made their marks as partners. Marie and Pierre Curie, James Watson and Francis Crick, Gertrude Elion and George Hitchings, and Otto Diels and Kurt Alder are among the most illustrious scientific duos. Michael S. Brown and Joseph L. Goldstein added their names to the list by discovering the LDL receptor in 1973.

From 1966 to 1968, Brown and Goldstein both attended Massachusetts General Hospital in Boston as interns and residents and became friends. Little did they know that their friendship and career would henceforth be intimately intertwined. From 1968 to 1971, they both spent two years at the NIH, Brown with Earl R. Stadtman, a distinguished biochemist, and Goldstein with Marshall W. Nirenberg, the 1968 Nobel laureate in physiology or medicine. After long discussions of science, the two close friends would relax by playing bridge, becoming great partners in the game.

In 1971 and 1972, respectively, Brown and Goldstein joined the Department of Internal Medicine at the University of Texas Southwestern Medical Center in Dallas. The department chair, Donald W. Seldin, inspired both of them to pursue a career in academic medicine. Initially, Brown and Goldstein maintained separate research laboratories, but in 1974, their laboratories formally merged.

Figure 1.9 Michael S. Brown © Nobel Foundation.

From 1974, Brown and Goldstein together studied the biology of familial hypercholesterolemia (FH), a genetic disease. FH was first described by Carl Müller, a Norwegian professor of internal medicine, as an "inborn error of metabolism" that produced high blood-cholesterol levels and heart attacks in young people. In the mid-1960s and early 1970s, FH was discovered to exist in two forms. The *heterozygous* form is less severe, striking 1 in 500 FH patients. They have cholesterol levels of 300–500 mg/dL and begin to have heart attacks at 30–40 years of age. The other, *homozygous* form is much more severe. Fortunately, homozygous FH is rare, striking only one in a million people. Homozygous FH patients, more prevalent in Quebec, Lebanon, and South Africa, have cholesterol levels greater than 1,000 mg/dL at birth and often begin to have heart attacks as children. These homozygous FH children have an average life expectancy of 14 years. Since it is at the upper extreme of the hypercholesterolemia spectrum, FH is an "ideal" form for scientists to study to understand the impact of cholesterol on coronary heart disease.

In order to decipher how cholesterol worked, a viable animal model bearing resemblance to FH was needed—after all, one could not just use people with this genetic defect as human guinea pigs! Serendipitously, in 1973, Japanese veterinarian Yoshio Watanabe at Kobe University (Kobe also produces the world's most expensive beef, Kobe beef, which is loaded with cholesterol) observed that a male rabbit in his colony had 10 times the normal concentration of cholesterol in its blood.[24] By appropriate breeding, Watanabe produced a unique strain of rabbits with high cholesterol levels. These rabbits promptly developed coronary heart disease and could serve as an optimal animal model for studying human FH.[25]

Interestingly, Brown and Goldstein started with an incorrect hypothesis. They initially thought that an enzyme had gone wild in FH patients and was producing the excessive cholesterol. For their studies, they focused on cultured skin cells from FH patients because liver samples were difficult to obtain, for obvious reasons. At one point, a doctor in Denver obtained a liver from an FH patient who had just undergone a liver transplant operation. Unfortunately, Brown and Goldstein had been working with skin cells, which behave quite differently from liver cells; they asked the Denver doctor to send them some skin samples rather than the liver.[26]

With the aid of Watanabe's rabbit model, they discovered that an enzyme was not the culprit. Instead, they found that FH patients lacked functional cell surface receptors for LDL, receptors that were responsible for removing cholesterol from the blood.[27]

Figure 1.10 Paul Ehrlich © Nobel Foundation.

In 1898, Paul Ehrlich was the first to propose the concept of *receptor*. He postulated a receptor as "receptive side chain." He drew the cell surface membrane protein that he later called "receptors." It was his idea that bioactive extracellular molecules were attracted to, and bound specifically to, these entities. Ehrlich's receptor concept was similar to Emil Fischer's "lock and key" hypothesis. In modern biology, the two concepts merged into the "ligand and receptor" concept. Today, we consider a receptor to be a molecule, such as a protein, that interacts with and subsequently binds to a ligand, such as a drug.

In addition, Brown and Goldstein also investigated how the liver processed cholesterol. They identified the key biochemical steps involving an enzyme called 3-hydroxy-3-methylglutaryl coenzyme A (HMG-CoA) reductase in the regulation of cholesterol through LDL receptors. They received the Nobel Prize in 1985 for their discoveries concerning the regulation of cholesterol metabolism. Later, they also demonstrated that statins could dramatically reduce levels of LDL cholesterol, the "bad

Figure 1.11 Ehrlich's sketch of a receptor (c. 1898).

cholesterol." Goldstein's speech at the Nobel banquet was probably the best summary of their synergy:

> Michael Brown and I are grateful to the Swedish academic community for bestowing this honor on us. Some may think that we are too young for this award. But let me point out that we work as a team. If our efforts were only additive, our combined age would be 45 *plus* 44 or 89 years. But our efforts are *more* than additive: they are synergistic. They have a multiplying effect. Our true collaborative age is 45 *times* 44 or 1980 years—surely old enough for a Nobel Prize. Our collaboration will continue tonight and we'll divide our allotted time in half.[28]

Recently, Lipids Research Clinics Coronary Primary Prevention Trial, (LRC-CPPT) a landmark 10-year study sponsored by the National Heart, Lung and Blood Institute, showed conclusively for the first time that reducing LDL cholesterol and total blood cholesterol can reduce the incidence of coronary heart disease and heart attacks in high-risk patients with significant amounts of plasma cholesterol.

Gradually, the experimental, genetic, and epidemiologic evidence all pointed to cholesterol as a major risk factor for cardiovascular disease. As the Framingham Heart Study revealed, other risk factors for coronary heart disease include smoking, a sedentary lifestyle, hypertension, and diabetes. Few medical questions have received more scientific attention than the correlation between cholesterol and coronary heart disease. The stage was now set for the discovery of drugs to lower cholesterol. "The lower, the better" has become a mantra for cardiologists all over the world with regard to LDL-cholesterol levels. For chronic and high-risk patients, the American Heart Association recommends LDL cholesterol levels be lower than 70 mg/dL, although normally 100 mg/dL is recommended for healthy individuals. Lowering cholesterol to prevent heart disease is the best form of medicine. Instead of "treating" a disease, this approach actually "prevents" a disease before it strikes, by eliminating a major risk factor. It is a cardiologist's dream come true.

In 1985, the National Heart, Lung, and Blood Institute announced the establishment of the National Cholesterol Education Program (NCEP). Despite a lack of fanfare for its creation, the NCEP was charged with the daunting task of reducing the prevalence of elevated blood cholesterol in the United States and thereby contributing to a reduction of coronary heart disease morbidity and mortality. Until 2004, an LDL-cholesterol level below 130 mg/dL was considered low enough. Then, the NCEP updated its guidelines, recommending that high-risk patients reduce their level even more—to less than 100—while patients at very high risk are given the "option" of reducing LDL cholesterol to less than 70 mg/dL. Patients often have to take more than one cholesterol-lowering drug to achieve these targets.[29]

There are currently four classes of viable cholesterol-lowering medications: nicotinic acid (niacin), bile acid sequestrants (resins), fibrates, and statins. Niacin came first in 1955. Bile acid sequestrants, such as cholestyramine, became available in 1957. Fibrates entered the picture in 1958. The most encouraging development in the treatment of hypercholesterolemia has been the introduction of statins in the late 1980s. By now, the therapeutic evidence, largely aided by the availability of statins, strongly suggests that lowering LDL cholesterol is tremendously beneficial in reducing the risk of having a heart attack or stroke.

The genesis of statins traces back to Akira Endo in Japan in the 1970s.

CHAPTER 2

Genesis of Statins

Early Cholesterol Drugs

As evidence grew that high blood cholesterol levels were linked to heart disease, scientists in both academia and industry began to look for drugs to lower cholesterol as early as the 1950s. Before Akira Endo discovered the first statin, mevastatin, in the 1970s, many things, including hormones, vitamins, and resins, were tried to lower cholesterol. Some worked, and some did not.

Thyroid hormone was one of the first drugs used for that purpose. The cholesterol-lowering properties of *dextro*-thyroxine were discovered by serendipity. At one point, surgical removal of part of the thyroid gland had been used to relieve angina, the pain brought on by exercise in coronary artery disease. Doctors observed that thyroid removal also raised the blood cholesterol level, which in turn sped up arterial degeneration. By deduction, the doctors reasoned that taking thyroid hormone should then decrease blood cholesterol levels. Initial clinical trials proved this theory, and *dextro*-thyroxine was used to lower cholesterol beginning in the 1950s, when thyroid extract became a standard treatment for hypercholesterolemic (high cholesterol) patients. Unfortunately, too much thyroid hormone made patients tremble all the time. Later, a large-scale, long-term clinical trial named the "Coronary Drug Project" established the association of *dextro*-thyroxine with ischemic heart disease as a severe side effect in men. As a consequence, thyroid hormone treatment was discontinued.

Women, in contrast to men, enjoy natural cardiac protection through the action of the female sex hormones, the estrogens. In 1930, a minute

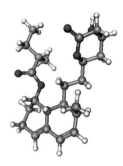

Figure 2.1 Molecular structure of mevastatin.

quantity of estrogen was isolated from the ovaries of 80,000 sows. In the 1950s, reports appeared that estrogen could lower blood cholesterol levels even more effectively than nicotinic acid, another anticholesterol drug used at the time. Unfortunately, men on estrogen for too long began to develop feminine traits, including breast enlargement and loss of libido, and other side effects, although they did acquire relative immunity from heart attacks until late in life. Due to the lack of safe and efficacious drugs, some doctors seemed willing to take their chances with estrogens. The April 27, 1962, issue of *Time* magazine reported that Dr. Jeremiah Stamler in Chicago saw good results using Premarin (an estrogen-containing drug isolated from pregnant mares) in high doses (5 mg/day) within three months after a patient's first heart attack. On the other hand, Dr. Jessie Marmorston in Los Angeles reported that he observed good results without ever going over 1.25 mg/day and that on this small dose his patients were not noticeably feminized. Dr. Stamler insisted that higher doses were necessary and that some feminizing was "unfortunately unavoidable." As one could imagine, despite Dr. Stamler's cavalier attitude, most men understandably shied away from estrogen hormone.

Nicotinic acid

Before the emergence of the statins, three classes of viable anticholesterol drugs existed: nicotinic acid (niacin), bile acid resins, and fibrates. Nicotinic acid (niacin) was first synthesized by Alfred Ladenburg at the University of Kiel, Germany, in 1897. In 1937, Conrad Elvejhem at the University of Wisconsin discovered its properties as a vitamin, now designated as vitamin B_3. He used nicotinic acid to cure the canine ailment "black tongue," a result of vitamin B_3 deficiency. When taken at the low, vitamin-level dose in milligrams, there was little change in the patients' cholesterol levels. In order to efficiently lower cholesterol, much larger doses were required. The cholesterol-lowering properties of gram doses of nicotinic acid were discovered in 1955 by Canadian pathologist Rudolf Altschul in the Department of Gerontology at the University of Saskatchewan in Saskatoon.

In the early 1950s, Altschul began to look for ways to decrease serum cholesterol levels in an attempt to inhibit experimental atherosclerosis in rabbits and in patients with degenerative vascular diseases. He fed various doses of nicotinic acid to rabbits and observed a decrease in serum cholesterol in 20 out of 24 tests. Meanwhile, Drs. Abram Hoffer and J. D. Stephen at the Pathology Department of Regina General Hospital in Regina, Canada had been studying the effect of large doses of nicotinic acid in the treatment of various diseases, especially schizophrenia. Altschul, Hoffer, and Stephen combined their interests and carried out an investigation of the influence of nicotinic acid on serum cholesterol in humans. They recruited 11 healthy volunteers with cholesterol levels of less than 250 mg/dL and 57 patients with cholesterol levels greater than 250 mg/dL and gave them daily doses of 1–4 grams of nicotinic acid. In 1955, they reported in the *Archives of Biochemistry and Biophysics* that nicotinic acid in gram doses lowered plasma cholesterol in normal subjects (6% decrease) as well as hypercholesterolemic subjects (22% decrease).[1] One year later, W. B. Parson and coworkers at the Mayo Clinic also reported on the effect of nicotinic acid. After administering 3 grams daily for 12 weeks, they observed an average 16% decrease of blood cholesterol for seven patients with familial hypercholesterolemia.[2] In the same year, massive doses of nicotinic acid were first used widely in Germany for lowering cholesterol. At that time, nicotinic acid was the only effective drug available for that purpose. Since then, nicotinic acid has established itself as a weapon in a doctor's arsenal to treat people with high cholesterol levels.

How does nicotinic acid work in lowering cholesterol? When Altschul, Hoffer, and Stephen published their landmark article, they had little clue as to nicotinic acid's mechanism of action. They incorrectly proposed that nicotinic acid promoted oxidative activities in the body and that the products, "oxycholesterols," were more readily excreted than pure cholesterol. The correct mechanism was actually not unraveled until very recently. In 2001, Anna Lorenzen and colleagues at the Heidelberg University in Germany reported that nicotinic acid works through binding to a G-protein–coupled receptor that they christened the "nicotinic acid receptor."[3] This receptor is present not only in adipocytes (fat cells) but also in the spleen and is responsible for lowering LDL-cholesterol levels.[4] Strangely, the body itself does not produce nicotinic acid, but must obtain it from external sources. Moreover, all existing nicotinic acid analogs, including Upjohn's acipimox, are less potent than nicotinic acid itself as agonists of the nicotinic acid receptor. Nicotinic acid is also the most effective of all its analogs at boosting the levels of HDL cholesterol.

Nicotinic acid is considered a drug, not a vitamin, when given at high enough doses for significant reduction of serum cholesterol concentrations. It is also the cheapest among all cholesterol-lowering drugs. In addition to lowering LDL and triglyceride levels, it also boosts the levels of HDL cholesterol, the good cholesterol. In 2005, nicotinic acid celebrated its 50th anniversary as a broad-spectrum lipid drug. Regrettably, the usefulness of nicotinic acid has been limited by its side effects, particularly intense flushing involving the face and the upper part of the body, which happens to many patients. Lately, nicotinic acid has experienced a reincarnation through formulations possessing reduced side effects.[5] Today, niacin can be bought in most vitamin and dietary supplement stores.

At the end of 2005, Merck surprised the medical world by announcing that it successfully completed the phase II clinical trials of MK-524a (trade name Cordaptive) and MK-524b. Cordaptive is a novel formulation of niacin that contains the antiflushing molecule laropiprant, and MK-524b is a combination drug containing both Cordaptive and Zocor, Merck's blockbuster statin drug. Merck discovered the antiflushing molecule laropiprant during its research into possible treatments for allergy sufferers. Laropiprant, a prostaglandin D_2 receptor antagonist, blocks prostaglandin D_2, the major cyclooxygenase metabolite of the arachidonic acid pathway. Laropiprant was later found to be central to counteracting niacin flushing. The large-scale phase III clinical trials for these two cholesterol-lowering drugs began in early 2006, involving 20,000 patients, and Merck submitted its results to the Food and Drug Administration (FDA) for approval in April 2008.[5] It was extraordinary that Merck tweaked niacin, a well-known vitamin that raises HDL in large doses, and tried to turn it into a blockbuster drug—Cordaptive could potentially fetch $2 billion in yearly sales. Sadly, on April 28, 2008, the FDA issued a "nonapprovable" letter to Merck with regard to Cordaptive, signaling the higher bar the agency now has for cholesterol drugs, considering that so many of them are now available, some in generic forms.

Resins

Cholestyramine, a quaternary ammonium ion exchange resin, was initially developed by Dow Chemical Company in Midland, Michigan, as a water softener. Impurities in the water bind to the resin's particles and can be precipitated out. In 1965, Sami A. Hashim and Theodore B. van Itallie of Columbia University reported that cholestyramine reduced cholesterol ester levels in a clinical study.[6] They gave patients with high cholesterol levels a total of 13.3 g of cholestyramine resin each day in four doses. The patients' serum cholesterol and cholesterol ester levels were reduced 50–80%. This was a resounding success, although that kind of phenomenal reduction of cholesterol has not been consistently reproduced. Hashim and van Itallie's report generated much excitement in cardiology. Merck, Sharp & Dohme tried to develop a drug for lowering cholesterol using cholestyramine, but the problem was that cholestyramine smelled like decaying fish and tasted little better. In 1967, working with Mead Johnson Laboratories (now a division of Bristol-Myers Squibb Company), Dr. Robert L. Fuson of Duke University added an orange flavor to the resin, which made the drug more palatable. One year later, cholestyramine was employed nationwide in the United States for decreasing blood cholesterol.

The year 1985 marked the establishment of the National Cholesterol Education Program (NCEP). The NCEP's treatment guideline placed cholestyramine and colestipol, a similar resin sold by Upjohn under the trade name Colestid,[7] as the drugs of choice for lipid lowering because they had been used for such a long time with a good safety record. At the time, the coordinator of the NCEP, James I. Cleeman, estimated that one out of five individuals referred for treatment would be placed on drug therapy, which was later proven to be correct.

How do these resins work? They are known as bile acid sequestrants, or bile acid binding resins. They form insoluble complexes with bile acids and are excreted in the feces, resulting in the elimination of bile acids. Cholesterol is a precursor of bile acid synthesis. As the body loses bile acids, it converts cholesterol in the blood to bile acid, thus lowering serum cholesterol. The end result is that the resins indirectly reduce cholesterol by forcing the body to draw on its cholesterol supply to make more bile acids. After Michael S. Brown and Joseph L. Goldstein's discovery of the LDL receptor, cholestyramine was found to work by increasing the number of LDL receptors, which in turn eliminate more of the LDL cholesterol. Since cholestyramine produces only a 15–20% increase in the synthesis of LDL receptors, there often is only a 15–20% drop in plasma LDL-cholesterol levels (far less impressive than Hashim and van Itallie's initial spectacular 50–80% reduction of cholesterol levels), which is not enough for most patients. Because of its low market potential after cholestyramine proved only marginally effective, Merck sold its rights to cholestyramine to Bristol-Myers Squibb, which marketed it under the trade name Questran. In July 1994, the Upjohn Company received clearance from the FDA to market its cholesterol-lowering Colestid tablets,[8] which gave patients a convenient alternative to the granulated medications on the market that need to be mixed with juice or water. The tablets were a prescription medication, used in addition to diet, to lower high total cholesterol and LDL cholesterol.

A conspicuous adverse effect of bile acid binding resins is constipation. When used in higher doses, depletion of fat-soluble vitamins, particularly A, D, and K, also occurs. Another drawback of these resins is that they taste like sand, so they are normally taken with fruit juice or other beverages to make them more palatable. A British physician commented on the original version of cholestyramine, called Cuemid: "It smelt like the bottom of a parrot's cage!" To increase patient compliance, Warner-Lambert Company infused cholestyramine into a "candy bar" formulation, which they sold under the trade name Cholybar.

Interestingly, in 1989, when Warner-Lambert received a shipment of cholestyramine as the active pharmaceutical ingredient for Cholybar from the chemical company Rohm & Haas, they discovered that the resin was contaminated. The launch of Cholybar was thus delayed, and Warner-Lambert filed a lawsuit against Rohm & Haas, asking for compensation.

Fibrates

In 1954, Imperial Chemical Industries Ltd. (ICI) in England discovered that certain plant hormone analogs decreased blood cholesterol levels. Intrigued by this chance observation, ICI screened all of the plant sterols that they could lay their hands on. Not surprisingly, most plant sterols were ineffective at lowering cholesterol, but they hit a jackpot when they discovered that clofibrate, though not a plant sterol, had a high level of cholesterol-lowering activity. ICI sold clofibrate under the trade name Atromid-S in 1958. In 1962, ICI's J. M. Thorp and W. S. Waring published a paper in *Nature* summarizing clofibrate's effects on lipids.[9] The article generated a flurry of research in the drug industry, resulting in many second- and third-generation "me-too" drugs, forming a family of fibrates that are more potent and safer than clofibrate.

Parke-Davis's gemfibrozil (trade name Lopid) was the second fibrate on the market. In order to find safer analogs of clofibrate, Parke-Davis screened more than 8,000 compounds similar to clofibrate—a gigantic feat at the time, considering that high-throughput screening was not yet available and the screening was carried out using animals![10] Gemfibrozil was discovered by Paul L. Creger in 1969 and was launched in 1982 with the trade name Lopid. Within the first year of launch, Lopid controlled about 40% of the $190 million anticholesterol business. Because Lopid was safer than clofibrate, it was widely used. When Mevacor, the first statin prescribed to patients, became available in 1987, Lopid was the most widely prescribed cholesterol-lowering agent at the time. Its annual sales peaked at $474 million in 1992. Abbott's Tricor, whose generic name is fenofibrate, is also a clofibrate analog.

Fibrates such as clofibrate (Atromid-S by ICI), gemfibrozil (Lopid by Parke-Davis), and fenofibrate (Tricor by Abbott) are primarily effective in lowering triglycerides by breaking down the particles that make triglycerides. These triglycerides are then used by the body in other ways. Lowering triglycerides can then lead to increased levels of HDL cholesterol, which is the most advantageous feature of the fibrates. Recently, fibrates have been found to activate a receptor essential for cholesterol and lipid metabolism (PPARα, peroxisome proliferator–activated receptor-α), and they have been used in treating type II diabetes since the late 1990s. As a result of the widespread use of clofibrate, scientists have even detected clofibric acid, a metabolite of clofibrate, in groundwater.

From 1964 to 1972, the World Health Organization investigated the effectiveness of clofibrate in more than 10,000 men between the ages of 30 and 59 who were at risk of a heart attack. Unfortunately, clofibrate was only found to decrease the incidence of nonfatal heart attacks in patients who did *not* have a history of heart disease. Since then, the FDA has restricted its use to limited indications. Recent studies have also shown that risk of drug–drug interactions increases 1,400-fold if statins are combined with fibrates. Therefore, fibrates should not be taken with statins.

I Refuse to Sign This Report!

In 1937, John Robson and Alexander Schonberg at Edinburgh University in Scotland found that triphenylethylene had estrogen-like activity. Seven years later, ICI's Frederick Basford discovered chlorotrianisene, a drug similar to triphenylethylene, as an estrogen-like drug. However, when chlorotrianisene was used for men as an anticholesterol drug, it caused breast enlargement and a decrease in libido like estrogen. Needless to say, finding a drug lacking feminizing side effects but directly lowering blood cholesterol was preferable, which was exactly what the William S. Merrell Company did.

Merrell was a drug firm located in Cincinnati, Ohio. Its parent company, the Vick Chemical Company, was a Fortune 500 company. In the 1950s, Merrell embarked on an ambitious program to find a cholesterol-lowering drug. Having learned much from the lessons of estrogen and chlorotrianisene, scientists in the company began to search for a molecule that possessed cholesterol-lowering activity without estrogenic side effects. They carried out extensive chemical modifications of chlorotrianisene and tested hundreds of compounds in animals. In 1959, Robert E. Allen, Frank P. Palopoli, and their colleagues patented triparanol as an anticholesterol drug.[11] Because the genesis of triparanol traced its roots back to estrogen, Merrell scientists jokingly called it a "nonestrogenic estrogen." The same year, Merrell sponsored a symposium at Princeton University, where triparanol garnered a favorable reception from the medical community. In April 1960, Merrell received approval from the FDA and in June began to market triparanol under the trade name MER/29. In the same year, Vick Chemical Company changed the name William S. Merrell Co. to Richardson-Merrell Inc.

Triparanol worked by blocking a late stage of cholesterol synthesis in the liver, forcing it instead to produce desmosterol.[12] As a result, unusually large amounts of desmosterol were left sloshing around in the blood. Clinicians initially had misgivings about the accumulation of large quantities of desmosterol, but Richardson-Merrell reported that they did not see serious adverse effects in animals, and few side effects were observed during clinical trials involving 2,000 patients. They claimed that triparanol was "the first safe agent to inhibit body-produced cholesterol" and "the first to lower excess cholesterol levels in both tissue and serum, irrespective of diet."[13] They also claimed that a single capsule daily would drop the blood cholesterol of 80% of patients to near normal, and presumably safer, levels.

Unfortunately, soon after triparanol reached a large number of patients, reports began to surface of side effects, including vomiting, nausea, hair loss, and compromised vision. When 18 patients in Rochester, Minnesota, were given the drug for three months, doctors found that, although the blood cholesterol was down by 13% on average, six of the patients had experienced hair loss and one had developed a skin disease.

Meanwhile, Merck, Sharp & Dohme also began searching for its own cholesterol-lowering drug. At the beginning of 1961, Merck used triparanol as a reference compound in their animal studies and reported that cataracts developed in the eyes of cats and dogs. It was suspected that cataracts were formed by deposition of nonmetabolizable steroids such as desmosterol in the cornea. Meanwhile, scientists at Upjohn in Kalamazoo, Michigan, reported similar results to Richardson-Merrell. On hearing the news, Richardson-Merrell initially suggested that the triparanol sample that Merck synthesized was not pure and that it was in fact the impurities that were causing the eye damage. Later, Richardson-Merrell claimed that the eye changes seen at Merck might be due to infection, because Richardson-Merrell "did not" observe any eye abnormalities.[13]

Mrs. Beulah Jordan at Richardson-Merrell was the laboratory technician who carried out triparanol's safety study on eight monkeys. In May 1959, the company asked her to falsify data to imply that the safety experiment was two months longer than it actually was. In addition, one monkey that was getting terribly sick mysteriously disappeared. Yet the data for that monkey appeared without any reported ill effects on the report submitted to the FDA for approval. Jordan refused to sign the falsified report and was so shaken by her experience that she quit her job at Richardson-Merrell. The FDA later learned from Jordan's husband that these cataracts had indeed been noted in animal studies. In April 1962, a team of FDA inspectors descended on Richardson-Merrell's Cincinnati headquarters. Two days later, the company withdrew triparanol from the market.[12] "Out of an abundance of caution," read the letter from Richardson-Merrell to 230,000 U.S. physicians, "we have determined that the sale of triparanol should be discontinued until all possible controversy is put to rest." Richardson-Merrell eventually admitted many cases of baldness, change of hair color, loss of body hair, and skin reactions ranging from dryness and itching to peeling and development of a fish-scale texture. In a few cases, triparanol was even suspected of reducing the body's protective white blood cells. By the time this all transpired, triparanol had been on the market for more than two years, and about 400,000 Americans had taken it, with hundreds of people experiencing unpleasant side effects.

The U.S. Justice Department eventually brought charges against Richardson-Merrell, including two scientists and one executive of the company, on 12 counts of supplying the FDA with "false, fictitious and fraudulent" data. After first pleading not guilty, pharmacologist E. F. van Maanen, laboratory chief William King, and vice president and director of research Harold Werner switched their pleas to *nolo contendere* (no contest) on eight counts. After federal judge Matthew M. McGuire made sure that each defendant understood that the new plea was "tantamount to a plea of guilty," the Justice Department dropped the four remaining counts.[14] Ultimately, Richardson-Merrell paid $200 million to 500 civil litigants, with an additional $80,000 fine. Ironically, Richardson-Merrell was also the American licensee of the most notorious drug of all time, thalidomide, from the German drug firm Chemie Grünethal. Tragically, thousands of babies whose mothers took thalidomide were born without arms or legs, instead having tiny, useless, seal-like flippers. The thalidomide episode was one of the darkest times in medical history. Fortunately, the United States was largely spared, thanks to Frances Kelsey, an FDA officer in charge of examining thalidomide, who withheld approval due to concerns about its safety.[14]

All in all, these three types of early cholesterol-lowering drugs—nicotinic acid (niacin), bile acid resins, and fibrates—still fell short in terms of efficacy and safety. One survey of three different drugs for 10,000 patients with elevated cholesterol levels found a lower incidence of cardiovascular events: for cholestyramine, 8% lower; for Lopid, 10%, and for statins, 20%. Clearly, statins are far superior to resins and fibrates.

Birth of Statins

The story of statins began with Dr. Akira Endo's obsession with microbes. Statins are cholesterol-lowering drugs that work by inhibiting HMG-CoA (3-<u>h</u>ydroxy-3-<u>m</u>ethylglutaryl <u>co</u>enzyme <u>A</u>) reductase, a key enzyme for the synthesis of cholesterol in liver.

Akira Endo discovered the first statin, mevastatin, in 1973 while working at Tokyo-based pharmaceutical company Sankyo Company in Japan. Little did he know that he had discovered the first of a class of molecules that would become the biggest moneymaker to date for the pharmaceutical industry—nearly $25 billion a year by 2005.

Endo's discovery of the first statin

Akira Endo was born in 1933 on a farm in the snowy north of Japan. When he was a boy, his grandfather taught him about the fungi that grew there. He was particularly fascinated by one poisonous mushroom that kills flies but not people, marveling that a natural substance could have such a subtle effect.[15] After obtaining his B.S. in Agriculture Sciences at Tohoku University in 1957, Endo joined the Fermentation Research Laboratories of Sankyo Pharmaceuticals. In 1965, fascinated by several excellent reviews on cholesterol biosynthesis by Konrad Bloch of Harvard University, Endo applied to Bloch for a postdoctoral fellow position in mid-December. Since Bloch had just been awarded the Nobel Prize in Chemistry a year before and was flooded with postdoctoral applications, he did not have any openings for the year 1966–1967. Endo then wrote a letter to P. Roy Vagelos, who was doing breakthrough research on the biosynthesis of fatty acids, at the National Institutes of Health. Endo did not receive an answer for four weeks. He later commented that if he had heard from Vagelos sooner, the first statin might have materialized a decade earlier. In late January 1966, Endo ended up taking a two-year appointment at the Albert Einstein College of Medicine in New York under the guidance of Professor Lawrence Rothfield to study phospholipid-requiring bacterial enzymes.[15] At the time, cholesterol was a hot area for ambitious scientists, and the U.S. press was reporting evidence that it played a role in heart disease. From Rothfield, who had worked as a practicing physician for nearly 10 years in the New York University Hospital before joining Albert Einstein College, Endo learned that hypercholesterolemia was a major risk factor for coronary heart disease. Moreover, it was the number one killer in Western countries. Endo's experience in the United States gave him the opportunity to learn about the mechanism of cholesterol regulation, which would prove invaluable in his later discovery of the first statin.

After completing his fellowship, Endo returned to the Fermentation Research Laboratories at Sankyo in 1968. Two years later, he hypothesized that levels of plasma cholesterol could be effectively decreased by inhibiting hepatic HMG-CoA reductase, an enzyme that promotes the rate-determining (slowest) step in cholesterol biosynthesis. His theory was actually very straightforward. During evolution, some fungi such as mushrooms developed a defense mechanism in which they produce a substance (called secondary metabolites in microbiology) that blocks the biosynthesis of cholesterol. Since bacteria require cholesterol-like compounds to grow, these substances could fend off invading bacteria by stifling the bacteria's ability to generate these compounds. In short, if you blocked HMG-CoA reductase, the slowest step of cholesterol production, you had a good chance of killing the other invading microbes.

In April 1971, Endo persuaded his superiors to dedicate resources to look for HMG-CoA reductase inhibitors in microbes. The company created a research unit to isolate such substances and assigned Dr. Masao Kuroda, a young chemist who had just joined Sankyo, and two assistants to work with Endo. They initially focused on fungi and mushrooms as sources of the secondary metabolites.[16]

For more than two years, Endo and his team worked long hours at their laboratory next to a train depot in southern Tokyo. "We were doing grunt work every day until we got sick of it," Endo later recalled.[16] During the screening process, the team did not just apply brute force, but also devised some clever procedures. For instance, at the time, HMG-CoA reductase was assayed principally by measuring the incorporation of carbon-14–labeled HMG-CoA into tritium (^{3}H) mevalonate. Because these radiolabeled samples were very expensive, the team improvised. Given that acetate is also a precursor of cholesterol synthesis, they searched for microbial culture broths that inhibited the incorporation of carbon-14–labeled acetate. Using the carbon-14–labeled acetate assay as the first filter, they then tested the active compounds at the next level involving more expensive ^{3}H-mevalonate.

Over a two-year period, Endo's team tested more than 6,000 microbial strains for their ability to block lipid synthesis. At first they isolated an antibiotic "citrinin" from the mold *Pythium ultimum*, but citrinin was an irreversible inhibitor of HMG-CoA reductase. Many irreversible inhibitors are toxic, and citrinin also fell into that group. Undeterred, Endo and his team persevered. In August 1973, one strain of fungus was finally found that produced an active compound, mevastatin. The fungus was *Penicillium citrium*, which belongs to the same genus of fungus that produced the antibiotic penicillin, *Penicillium notatum*. Although there are numerous species in the *Penicillium* genus, it seems profound that penicillins and statins, two of the most important classes of modern drugs, were first isolated from the same genus. It is also not surprising that some people call the statins "penicillin for the heart." In 1973, Brown and Goldstein reported that HMG-CoA reductase activity in cultured mammalian cells is suppressed by LDL but not by HDL. Their work on cholesterol metabolism strongly supported Endo's studies in both experimental techniques and the general idea of developing HMG-CoA reductase inhibitors.

Mevastatin, the active compound that Endo isolated, was the first statin. Endo's group began by extracting a whopping 600 liters of the culture filtrate and finally isolated only 23 mg of crystalline mevastatin.[17] In 1976, A. G. Brown and colleagues at Beecham Pharmaceutical Research Laboratories in England also isolated mevastatin from another *Penicillium* fungus, *Penicillium brevicompactum*. They named it compactin and elucidated its structure by X-ray crystallography, confirming that compactin and mevastatin were indeed the same molecule. Brown and coworkers initially discovered compactin as an antifungal agent, but abandoned it due to its low potency. They also tried to assess its usefulness as a cholesterol drug, but they used rats for the animal model and saw no plasma cholesterol change at all.

As we now know, mevastatin has a specific effect in lowering plasma LDL. Rats have hardly any LDL in plasma to begin with—nearly all of their cholesterol is contained in HDL—so it is not surprising that mevastatin did not lower cholesterol levels in rats.[19] This species of animal is a lousy model for lowering plasma cholesterol anyway since statins induce a massive amount of HMG-CoA reductase buildup in rat liver. When given multiple doses of mevastatin, the enzyme level sometimes rose 3–10 times compared to the control rats.

Endo discovered the same thing in early 1974. When he fed rats a diet supplemented with 0.1% mevastatin for 7 days, he observed no efficacy for lowering plasma cholesterol levels. The same thing happened for even higher doses after 5 weeks of treatment.

Just when Endo was scratching his head over mevastatin's lack of efficacy in rats, he ran into a colleague at a "watering hole" near the laboratories one night. This colleague offered some hens for testing, saying they were about to be sacrificed anyway.[17] Hen's eggs generally contain about 300 mg of cholesterol; two-thirds of it is obtained from the hen's diet, and the remainder is synthesized by the liver. At the beginning of 1976, Endo treated laying hens with 0.1% mevastatin for 30 days, and their cholesterol levels were reduced by as much as 50%! Excited by the outstanding results, Endo and coworkers then fed mevastatin to dogs and observed similar success. Sankyo had patented mevastatin in 1974, and Endo published his discovery two years later. In early 1977, with confidence mounting, they gave it to monkeys for 11 days, and saw a 21% reduction of plasma cholesterol at a lower dose (20 mg/kg) and a 36% reduction at a higher dose (50 mg/kg). Since monkeys are primates with physiology very similar to that of humans, Endo was confident enough to request that management take mevastatin into human clinical trials.

Unfortunately, Sankyo's managers did not share his enthusiasm. The odds of success in making a drug were, and still are, much lower for new mechanisms of action than developing those "me-too" drugs with an already proven mechanism. Strategically, the Sankyo brass played its odds and decided to focus their resources on developing refinements of then-existing cholesterol drugs such as fibrates.

In September 1977, Endo attended the 6th International Symposium on Drugs Affecting Lipid Metabolism in Philadelphia. Almost no one attended his talk on mevastatin because the field was mesmerized by fibrates and bile acid sequestrants. It now looked as though mevastatin was destined to fade into oblivion.[18]

The fate of mevastatin

Despite management's lack of enthusiasm, Endo garnered the support of his direct supervisor, Dr. M. Arima, to move forward with clinical trials in humans. Stealthily, Endo approached Dr. Akira Yamamoto at Osaka University, who had been treating patients with familial hypercholesterolemia (FH). In February 1978, Yamamoto proceeded to test mevastatin on a patient with FH. Nowadays, such an act would be unthinkable because testing a drug in humans requires approval by regulatory agencies, such as the FDA, via an Investigational New Drug application. Regrettably, Yamamoto gave an 18-year-old girl a high dose of mevastatin, and she experienced muscle weakness. Although his superiors at Osaka University ordered him to discontinue, Yamamoto secretly continued to give mevastatin at low doses to several other patients—a behavior that today would be unethical and would come with serious legal ramifications. Encouragingly, at lower doses, Yamamoto's FH patients saw 22–35% reduction in total cholesterol levels. When the positive data were presented to Sankyo management, they finally agreed to put mevastatin in a formal clinical trial, which began in more than 10 groups throughout Japan in early 1979 for patients with severe hypercholesterolemia. But by mid-1980, Sankyo halted the trials because long-term safety studies on dogs revealed that some of them started to show intestinal tumors at high doses (100–200 mg/kg, which was 500- to 1,000-fold higher than the effective dose). Mevastatin was completely safe to use at 25 mg/kg. Even today, Endo argues that it was a mistake to discontinue the clinical trials.[16–18]

In retrospect, at least four obstacles hindered the development of Endo's mevastatin. First of all, mevastatin was not very effective in rats and mice under normal feeding conditions. It turns out that an enzyme (cholesteryl-7-hydroxylase) was significantly suppressed by the administration of mevastatin, which resulted in false-negative results, but not until later was it learned that the rat was a poor model for cholesterol-induced atherosclerosis, a fact widely known nowadays. Even today, Nikolai Anitschkov's experiment has not been reproduced in rats, although it is successful in nearly all other animal models. Second, the appearance of fine crystals in rat hepatocytes made clinicians nervous about their ramifications. Third, an unwritten and underlying reason was the catastrophe of thalidomide, which caused thousands of deformed babies worldwide, including Japan. Doctors were therefore extremely wary of even the slightest hint of toxicity. The last straw was the intestinal tumors found in dogs, even though the intestinal lymphoma only appeared when extremely high doses of mevastatin (100–200 mg/kg) were given to dogs for a prolonged time (24 months).[16–18]

Despite mevastatin's failure to reach the market, not all was lost for Sankyo in the statin area. Sankyo reaped the fruits of their investment in Endo's effort through pravastatin, first isolated from the urine of dogs that had been fed mevastatin (pravastatin is a metabolite of mevastatin). Indeed, one difference between these two molecules is that pravastatin has an extra hydroxyl group, arising from an oxidation of mevastatin. The other difference is that pravastatin has a ring-opened dihydroxycarboxylic acid side chain, whereas mevastatin has a lactone ring. Still, due to the tumor scare from dogs, Sankyo did not gather enough courage to test pravastatin until May 1984. Five years later, Sankyo comarketed pravastatin with Bristol-Myers Squibb, which had the marketing rights for the American market. Bristol-Myers Squibb sold pravastatin under the trade name Pravachol following FDA approval in October 1991.

By the year 2001, six statins were on the market: Pravachol, along with Mevacor (lovastatin), Zocor (simvastatin), Lescol (fluvastatin), Lipitor (atorvastatin), and Baycol (cerivastatin). Competition was fierce. In order to improve its Pravachol's third-place position in the booming $10 billion market for cholesterol-lowering drugs, Bristol-Myers Squibb enlisted Hollywood celebrities in a campaign to educate the masses about cardiovascular disease. It tapped stars like Kirk Douglas, Sylvester Stallone, Dana Carvey, and Angela Bassett to help. As part of the campaign, celebrities like Douglas, who suffered a stroke in 1995, spoke about how heart disease had affected their lives. "Heart disease is the nation's No. 1 killer," said Douglas, then 84. "If people have more knowledge, they can better take care of themselves."[19] By April 2006, Bristol-Myers Squibb lost patent exclusivity on Pravachol, and now generic pravastatin is available in the U.S. pharmacies.

Endo's fate

In 1978, while mevastatin was still in the hands of clinicians, Endo's career at Sankyo came to an abrupt halt. Willingly or unwillingly, Endo had offended enough managers that he was forced out of Sankyo. Deprived of recognition for his groundbreaking discovery and his chances of advancement within the company, in December 1978 he decided to take a professorship in the Department of Applied Biological Science at Tokyo Noko University. It was a bleak scene during his move—the company even forbade colleagues in his laboratory from helping him carry his boxes of papers to the moving truck.[15]

Figure 2.2 Akira Endo, 2005.

On October 1, 2003, scientists around the world gathered in the Kyoto International Conference Hall for a grand symposium commemorating the 30th anniversary of the discovery of mevastatin. It was held in honor of Dr. Endo for opening the statin field. The symposium attendees all spoke rhapsodically of Endo's great achievement, and many believed that Endo deserved a Nobel Prize as the discoverer of the first statin. Brown and Goldstein also sent a tribute to the symposium: "The millions of people whose lives will be extended through statin therapy owe it all to Akira Endo and his search through fungal extracts at the Sankyo Co."[20]

In 2004, Dr. Endo went to see his physician, and tests showed that his total cholesterol level had reached 240 mg/dL, with an LDL cholesterol level of 155. Endo's doctor told him: "Don't worry! I know some very good drugs to decrease your cholesterol." He prescribed Mevacor for Endo, having no clue that he was treating the father of statins. Endo took Mevacor for a while but stopped because he chose to instead exercise and had brought his LDL cholesterol level down to 130.[21]

How do statins work?

In the early 1960s, Konrad Bloch delineated the biosynthesis of cholesterol—the process by which it is produced in the body. It goes through the acetic acid → squalene → cholesterol cascade. An early step, also the slowest and thus rate-limiting step, involves the reduction of HMG-CoA to mevalonate, which is subsequently transformed into cholesterol after several steps. This crucial reduction process is accomplished by an enzyme called HMG-CoA reductase, the rate-controlling enzyme in the biosynthetic pathway for cholesterol. Therefore, it is reasonable to assume that if one could block the function of HMG-CoA reductase, the assembly line for cholesterol production would be suppressed.

Endo determined mevastatin's biochemical mechanism of action in 1976. Like all statins, mevastatin is a potent competitive inhibitor of the HMG-CoA reductase. That statins specifically inhibit HMG-CoA reductase should not be underappreciated. Indeed, much of the proven safety associated with statins is due to their action on HMG-CoA reductase, which is relatively early in the cholesterol biosynthesis pathway. Contrasted with triparanol, for instance, their safety has been ably demonstrated in the last two decades.

For triparanol, the toxicity is *mechanism-based*. Triparanol blocks the last step of cholesterol biosynthesis, so the synthesis then diverts to squalene intermediates, giving rise to very large lipophilic compounds. These compounds are so greasy that the human body cannot remove them through metabolism, so they cause cataract formation. Conversely, statins block cholesterol biosynthesis by inhibiting HMG-CoA reductase at the

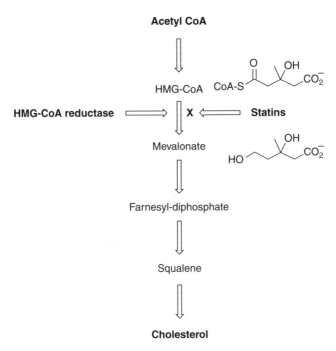

Figure 2.3 The role of HMG-CoA reductase and statins in the production of cholesterol.

early stage of the cascade. As a consequence, small molecules accumulate, not large, greasy molecules as with triparanol. These small molecules are innocuous and easily removed through metabolism. Therefore, toxicity for statins is largely *molecule-based*, although the latest scholarship points out that coenzyme Q10 has a lot to do with the toxicity. Statins are safe drugs in general, but each individual statin has its own safety profile. While Zocor and Lipitor are extremely safe, the unfavorable risk–benefit profile of Baycol resulted in its withdrawal from the market in 2003.

CHAPTER 3

Merck's Triumph

Merck was not the first to discover a statin (Akira Endo at Sankyo was), but it was the first company to bring one successfully to market. In September 1987, the FDA approved Merck's Mevacor (lovastatin) for marketing in the United States while Sankyo's mevastatin was resting in peace in the big graveyard of drugs that failed to reach the market.

Merck

Humble beginnings

In 2007, Merck & Co. was ranked the third largest pharmaceutical company in the world, both by capital and by revenue, behind only Pfizer and GlaxoSmithKline. The drug juggernaut traces its origin to a humble apothecary shop in Darmstadt, a central German city 20 miles south of Frankfurt.[1] In 1668, 47-year-old apothecary Friedrich Jacob Merck purchased the Engelapotheke (Angel Pharmacy) in the Hessian town and began trading fine chemicals. Most popular was laudanum, an alcoholic solution of opium, which was in every physician's medicine chest and used as a "panacea" for many illnesses. In 1827, Heinrich E. Merck inherited the family business and established a chemical laboratory named "E. Merck & Co." beside the Angel Pharmacy.[2] There, he pioneered the commercial large-scale production of various medicinal alkaloids, including veratrine, codeine, atropine, quinine, coniine, and morphine.[3]

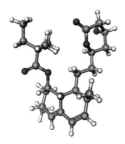

Figure 3.1 Molecular structure of Mevacor.

In 1860, Albert Niemann, working in Friedrich Wöhler's laboratory in Göttingen, Germany, isolated the active principle of coca leaves as a white crystalline alkaloid and christened it "cocaine." Since new compounds were routinely tasted, Wöhler recorded that "cocaine was a substance which had a somewhat bitter taste and exerted a numbing influence upon the gustatory nerve, so that they became almost completely insensitive."[4]

Capitalizing on its financial success with alkaloids, E. Merck began to isolate cocaine from coca leaves in the early 1880s and aggressively marketed it as a pain killer. Interestingly, Sigmund Freud, the father of psychoanalysis, enthusiastically took part in exploring cocaine's medical utilities. After procuring some cocaine from E. Merck, the young neurologist swallowed a small quantity of the drug, which calmed his stomach and boosted his libido. Freud also applied some cocaine locally to himself and found that it temporarily paralyzed the sensitivity of a certain area without any marked effect on the central nervous system. Working together, he and his friend Carl Köller discovered cocaine as a local anesthetic that has since been very important in minor surgery.

Another drug that made Merck "famous" was Ecstasy, also known as 3,4-methylenedioxymethamphetamine (MDMA). In 1912, E. Merck was looking for cardiovascular drugs and drugs that stop bleeding. In the process of synthesizing hydrastinine, Dr. Freund, a chemist at Merck, isolated MDMA as a by-product. At the time, it was just a nuisance during purification, but E. Merck nonetheless filed a patent for this series of compounds two years later.[5] Unfortunately, abuse of Ecstasy began to emerge in the 1980s, quickly replacing amphetamine and LSD to become one of the most abused drugs.

Nowadays, the original E. Merck & Co. still exists in Darmstadt but with the name Merck KGaA. However, in terms of capital, influence, and fame, Merck KGaA is decisively dwarfed by Merck & Co., formerly a unit of the German-based E. Merck and now an independent company.

George W. Merck

In 1887, E. Merck & Co. sent a brilliant chemist, Theodore Weicker, who had been with the firm for 10 years, to New York in an attempt to capitalize on America's newly acquired wealth. In 1891, 24-year-old George Merck, a grandson of Heinrich E. Merck, was sent to New York to oversee the newly established American office. Nowadays, Merck & Co. considers its foundation date as 1891. At first, George Merck simply imported whatever drugs E. Merck produced and sold them to Americans. But the high tariff that the U.S. government levied against imported goods prompted him to begin making those drugs in the United States. In February 1900, George Merck purchased a tract of 120 acres of New Jersey woodland in a small town named Rahway. There, Merck began manufacturing chloral hydrate, iodine, bismuth salts, acetanilide, narcotics, salicylates, and alkaloids. Not too surprisingly, George Merck and Theodore Weicker, two strong-willed men, engaged in an acrimonious power struggle, which ended with George Merck's victory. Embittered, Weicker sold his share of the business to George Merck in 1903 and, with another partner, bought another drug firm, E. R. Squibb & Sons, from the two sons of its founder.

World War I ended in November 1918 after five years of bloody conflict with the Allies, including France, America, England, and Russia, defeating Germany. In the Treaty of Versailles, the Allies imposed harsh terms upon Germany. The U.S. government also confiscated German companies, including the U.S. branches of Merck and Bayer, and auctioned them off to domestic bidders. Merck & Co., originally a foreign affiliate of E. Merck, has been an independent American company since then. On October 21, 1926, the company's founder, George Merck, died at his home in New Jersey at the age of 59. A year earlier, he had passed the control of the company to his son, George W. Merck, who would go on writing a splendid chapter of Merck's history.

George W. Merck, born in 1894, grew up in the pleasant countryside of Llewellyn Park, New Jersey. He attended Harvard University and graduated in 1915 with a bachelor's degree in chemistry, one year ahead of normal schedule. His dream of earning a chemistry Ph.D. in Germany was shattered by the outbreak of World War I. Instead, he went to work in his father's firm and took over the family business in 1926, when he was only 32 years old. He determined to transform the small fine chemical supplier into a premier drug firm. In 1933, Merck established the Merck Institute for Therapeutic Research and began its own pharmaceutical research. Quickly, Merck moved to the forefront of medical research in at least four areas: vitamins, sulfa drugs, antibiotics, and hormones.

George W. Merck stood an imposing six feet five inches tall and weighed 250 pounds. He was a staunch Republican and also served as chairman of the New Jersey Republican Party. Looking back at George W. Merck's remarkable life, his philosophy on drug discovery still holds true today: "For one thing, there is always serendipity. Remember the story of the Three Princes of Serendip who went out looking for treasure? They did not find what they were looking for, but they kept finding other things just as valuable. That's serendipity, and our business is full of it."[6]

While giving a speech at the Medical College of Virginia at Richmond in 1950, George W. Merck immortalized the Merck values that would resonate for decades to come: "We try to remember that medicines are for the patient. We try never to forget that medicine is for the people. It is not for profit. The profits follow and if we've remembered that, they have never failed to appear. The better we have remembered it, the larger they have been."[1]

Penicillin

During the reign of George W. Merck, the company achieved numerous successes in discovering and developing life-saving drugs. Penicillin, streptomycin, and cortisone were the crowning jewels of Merck's scientific achievements. In fact, Merck & Co. first made its name producing penicillin, which transformed Merck from a small fine chemical supplier into a premier drug company in America.

In the summer of 1928, while working in the Department of Morbid Anatomy at St. Mary's Hospital in London, 47-year-old Scottish bacteriologist Alexander Fleming left his petri dishes on the bench without cleaning them before his two-week vacation.[7] When he came back, he noticed that the colonies of one bacterium were dissolved around some thick greenish molds that had appeared. The green mold was later identified as *Penicillium notatum*, which had inadvertently blown into Fleming's culture dish from downstairs, where a professor was doing experiments with the mold. Fleming isolated crude penicillin extracts to test on rabbits and even washed one of his colleagues' infected eyes, with promising results and no significant toxicity. Unfortunately, his serendipitous discovery did not find acceptance in the medical community until 1938, when Australian pathologist Howard Florey and his biochemist colleague Ernst Chain decided to reexamine Fleming's penicillin. Working at the Sir William Dunn School of Pathology at Oxford University, Florey and Chain assembled a team of scientists and carried out extensive research to produce and purify the active principle from the *Penicillium notatum* mold.[8] The Oxford team isolated enough penicillin to test on Swiss albino mice, with spectacular results. They even tested it on an Oxford policeman whose mouth had become infected from a scratch while he was pruning roses. Although penicillin brought him back from the brink of death, he unfortunately died a few weeks later when the penicillin supply ran out. As the end of World War II seemed nowhere in sight, there was an urgent need to make enough penicillin to help those that were wounded in battle.

Pummeled by the Luftwaffe bombardment during the Battle of Britain, England was not a safe place to investigate and manufacture penicillin. Florey therefore turned to America for help. Under a $5,000 grant from the Rockefeller Foundation, in July 1941 Florey and his microbiologist colleague Norman Heatley boarded an airplane carrying the *Penicillium* mold and some penicillin samples in a briefcase. When they arrived in New York City, their first stop was the Rockefeller Foundation, where they discussed the penicillin situation with Alan Gregg, the head of the Foundation's Medical Sciences Division. Impressed by penicillin's therapeutic potential, Gregg recommended a direct negotiation with a drug firm under the auspices, and with the backing, of the government. Since Gregg knew George W. Merck personally and thought highly of him, he suggested that Florey contact Merck. In the end, in addition to Merck & Co., Charles Pfizer, E. R. Squibb & Sons, and Lederle Laboratories also took part in the initial foray into penicillin production. In 1942, after Heatley's initial nine-month stay at the Northern Research Laboratory of the U.S. Department of Agriculture in Peoria, Illinois, he was employed by Merck for six months to help with the mass production of penicillin.

George W. Merck allotted $100,000 to the penicillin project. He also made a decision to focus on the chemical synthesis of penicillin and assigned the job to Max Tishler, an organic chemist who was the head of Merck's process chemistry department. Tishler, born in 1906 in Boston, was the fifth of six children of Jewish parents who emigrated from Germany and Romania.[9] Because his father, a cobbler, had left the family when Max was five years old, Max had to work throughout his youth to support his family. Even in graduate school at Harvard, he sold candies to his labmates. After graduating with a Ph.D. in organic chemistry under

Elmer P. Kohler in 1934, he had difficulty finding a job and was told that "Jews have a hard job getting placed and you won't get anywhere."[10] Therefore, Tishler stayed at Harvard for another three years, teaching, doing research, and revising James Conant's textbook on organic chemistry. After being rejected by DuPont, he was hired by Merck in Rahway, New Jersey, in 1937. He loved to take new compounds and develop efficient processes for producing them on a large scale. When he led the penicillin project in 1941, Tishler, an archetypical type A personality with an extraordinary combination of energy and ability, pushed his chemists hard and himself even harder. Working on Saturdays and Sundays was nothing out of the ordinary for his team. Once, when a chemist requested more penicillin because he believed too little penicillin was allotted to him for research, Tishler told him, "Remember, when you are working with those 50 to 100 milligrams of penicillin, you are working with a human life."[11] On one occasion, the Merck process people were working to synthesize a large quantity of an arylhydrazone intermediate that happened to be brilliant red in color. Unfortunately, one day, a reactor broke during the process and a large amount of the compound spilled onto the floor of the plant. When Tishler was summoned, he took one look at the disaster on the floor and exclaimed: "That had better be blood!"[11]

Because penicillin was rapidly recognized as a wonder drug, the chemical synthesis of penicillin became the top priority for the Allies' chemists. More than 1,000 organic chemists in 39 major research laboratories took part in that project. In addition to Merck, participating American drug firms included Abbott, American Cyanamid, Eli Lilly, Parke-Davis, Pfizer, Hoffmann La Roche, Squibb, Upjohn, and Winthrop. The scale of effort was rivaled only by the Manhattan Project to build the atomic bomb.[11] Unfortunately, in the end, the chemical synthesis of penicillin was not accomplished during World War II. After the war, almost all abandoned their efforts, deeming it impossible. Only one group kept working on it: John C. Sheehan at MIT. Sheehan, who had worked on the penicillin synthesis under Tishler at Merck, completed the first total synthesis of penicillin V in 1957.[11] Tishler even aided Sheehan's pursuit by providing large quantities of intermediates prepared by the Merck process chemistry department.

Meanwhile, Merck developed a submerged fermentation process that sped up the production in its new plants, which were in full operation by 1943 and produced the first dose of penicillin in America. Ann Miller was the first American whose life was saved by penicillin produced by Merck. Miller, wife of the Yale athletics director, was on her deathbed from a streptococcal septicemia infection following a miscarriage. It took Tishler's chemists three days to purify just 5.5 grams of precious penicillin from the first batch of their fermentation process. The drug was then flown from Rahway, New Jersey, to New Haven, Connecticut. Miller was saved, and she lived to be 95 years old!

Merck's penicillin was also responsible for saving hundreds of burn victims from the Coconut Grove fire in Boston.[11] On November 28, 1942, more than 1,500 people were packed into the Coconut Grove Club when a fire erupted and killed 300 partygoers on the spot, badly burning hundreds more. The standard treatment consisted of coating the burn with ointments or tannic acid, but neither would prevent infection, so many were expected to die. When Tishler was informed of the situation, he arranged for his chemists to work around the clock to purify the crude penicillin broth. When one shift went home for the day, another shift came to work for the night. When the drug was finally ready, it was sent to Boston with great fanfare. According to the *Boston Globe*:

> Police escorts from four states accompanied a consignment of an as yet unnamed drug rushed to the Massachusetts General Hospital early this morning from the Merck & Co., Inc. Laboratory in Rahway, N. J. for treatment of fire victims. A 32-liter supply of this drug, described as priceless by a laboratory technician, will be used to prevent infection from the burns. (December 2, 1942)

At the end, Merck's penicillin saved many lives, and Merck was transformed from a small chemical firm into a well-known player in the pharmaceutical industry.

Cortisone

If penicillin was the drug that catapulted Merck's status to a bona fide drug firm, without a doubt, the synthesis of cortisone galvanized Merck into a research institute that rivaled the best in both academia and industry. In 1944, Lewis H. Sarett at Merck was the first to synthesize cortisone from bile acid, a steroid isolated from cattle bile.

In 1941, Philip S. Hench and Edward C. Kendall at the Mayo Clinic in Rochester, Minnesota, discovered that cortisone had strong antirheumatic properties. Meanwhile, cortisone also had become a top priority to the Allied military medical establishment. Rumors indicated that Luftwaffe fighter pilots were able to fly at unusually high altitudes without oxygen deprivation because they were being treated with cortisone. Understandably, in light of this military advantage, the Allies wanted to make their own cortisone to level the battlefield.

Sarett was born in 1917 in Champaign, Illinois. His father was a speech professor at the University of Illinois. He studied steroid chemistry under Everett S. Wallis at Princeton in 1940[12] and accepted an offer to work at Merck in 1942 after earning his Ph.D. in only two and a half years. When Sarett arrived at Merck, Tishler assigned him the cortisone synthesis and let Sheehan work on the penicillin synthesis. Years later, Sheehan and Sarett reminisced about what the outcome would have been if they had switched assignments; they agreed that both of them would have probably failed.[11]

Ironically, Sarett's initial assignment was to prove the existence of the C-11 oxygen on a molecule. He did not make much headway in the first several frustrating months. "You haven't made much progress, have you?" His supervisor commented to him: "Maybe Merck isn't the company for you."[13] This chilling suggestion sent Sarett into shock. He spent a nerve-racking weekend and proved that the oxygen was on the C-12 position rather than C-11 as claimed. After overcoming that initial obstacle and gaining back the confidence of his supervisor, Sarett began to tackle the challenging task of synthesizing cortisone. Without an assistant, Sarett single-handedly made 18 mg of cortisone from bile acid, although his brilliant synthesis was 37 steps long. Tishler did supply him with some of the intermediates required for his efforts in order to speed up the project. It took another two years for Tishler and his colleagues at Merck Process Development, especially Jacob van de Kamp, to further optimize Sarett's original synthesis. This effort finally paid off, though, and they prepared 100 grams of cortisone and distributed portions of this material to research groups around the country. In September 1948, Hench received 5 grams of cortisone from Merck. A dose of 50 mg and then 100 mg of cortisone was administered to a desperately ill 29-year-old Mrs. Gardner, who looked 50 and had not been able to get out of bed for the last five years without assistance because of her rheumatoid arthritis. After three days of cortisone injections, Mrs. Gardner miraculously recovered. She even went to downtown Rochester and treated herself to a three-hour shopping spree! This triumph was touted as one of the greatest achievements in medical history. Hench also thought very highly of Merck's achievements, writing that Merck was "writing a brilliant chapter in the history of pharmaceutical manufacturing, accomplishing the impossible."[14] Both Tishler and Sarett went on to become members of the National Academy of Sciences, a rare honor for industrial chemists.

Streptomycin

The discovery of streptomycin was also closely associated with Merck. Streptomycin was one of the first effective drugs against tuberculosis, an infectious disease that has plagued humanity for millennia.

Selman Abraham Waksman, an immigrant from a small Jewish town in the Ukraine, was a professor of soil microbiology at Rutgers. Inspired by his student René J. Dubos, he started to focus his attention on discovering antibiotics in soil. In 1938, Waksman and George W. Merck struck a deal resulting in the establishment of a Merck fellowship in fermentation studies in Waksman's laboratory. Under the agreement, Waksman received chemistry support from Merck and was able to use the extensive facilities that Merck could offer. In return, Merck received the patents on any process Waksman developed.[10]

In October 1943, Waksman's student Albert Schatz isolated an antibiotic that was very effective at killing gram-negative bacteria, which were not touched by penicillin. Waksman christened it streptomycin. With assistance from Merck for large-scale production and the Mayo Clinics for animal testing and clinical trials, streptomycin was proven to be both safe and effective in treating tuberculosis. Astonishingly, only three years elapsed from streptomycin's discovery to its use in the first successful treatment of a human patient.

Tishler led Merck's microbial group that developed the fermentation process for producing bulk quantities of streptomycin. At the

Figure 3.2 Selman Waksman
© Nobel Foundation.

beginning of its development, they experienced a recurring difficulty with the purification of streptomycin. After fermentation, streptomycin was purified by adsorption onto charcoal and then elution to liberate the material from the charcoal. However, the product was always contaminated with histamine and therefore the drug produced histamine-like allergic reactions in patients—elevated blood pressure, pain, and allergic rashes. Sheehan, a group leader at Merck in the 1940s, solved this problem overnight. Streptomycin, an amino sugar, is extremely soluble in water, but virtually insoluble in certain organic solvents that are themselves immiscible with water. Mimicking an old German process for purifying cane or beet sugar, Sheehan mixed the impure sample in a separatory funnel charged with water and phenol. The brown-colored histamine impurities immediately went into the phenol layer. After separating the phenol layer, the clear water solution was freeze-dried, furnishing the purest streptomycin they had yet produced.

The U.S. Patent Office granted U.S. Patent 2,449,866 titled "Streptomycin and Process of Preparation" to Schatz and Waksman on September 21, 1948. It has since become one of the top 10 patents that have changed the world. Rutgers licensed streptomycin to Merck. In an extraordinary gesture for a drug firm, George W. Merck, at the request of Waksman, signed away the exclusive right for developing streptomycin, allowing any drug company to make and sell streptomycin. Merck's magnanimous act greatly enhanced his company's reputation not only for its science, but also for its humanitarian fame. Years later, in 1987, Merck's then-CEO, P. Roy Vagelos, made a similar generous decision that Merck would donate Mectizan free to anyone in the world who needed it. Mectizan (ivermectin) is an antiparasite drug that is very effective at preventing river blindness, a disease transmitted by blackflies that has plagued sub-Saharan Africa.

As a result of competition among many drug firms, the price of streptomycin quickly became very affordable. To compensate for Merck's financial loss, Rutgers refunded the first half million dollars in royalties toward Merck's initial outlay in commercializing streptomycin. Waksman, who in 1952 won the Nobel Prize in Physiology or Medicine for the discovery of streptomycin, often said that without Merck, most, if not all, of the antibiotics that he isolated would have remained bibliographic curiosities.

Glorious history

The merger with Sharp & Dohme to create Merck Sharp & Dohme in 1953 established a solid foundation for a fully integrated, multinational producer and distributor of pharmaceutical products.

In the early 1950s, Merck chemist Frederick C. Novello wanted to make diuretic agents by synthesizing analogs of an older sulfa drug, dichlorophenamide.[15] Unexpectedly, one of his reactions gave a ring formation product rather than the linear derivatization product. The bicyclic ring formed was a benzothiadiazine derivative, which was later named chlorothiazide. Although disappointed by not getting what he had intended, Novello submitted the compound for screening anyway. It proved to be a potent diuretic without elevation of bicarbonate excretion, an undesired side effect. Chlorothiazide (trade name Diuril) was groundbreaking. It was the first ever nonmercurial, orally active, diuretic drug whose activity was not dependent on carbonic anhydrase inhibition. In 1957, George deStevens, a chemist at Ciba, transformed Merck's chlorothiazide into hydrochlorothiazide, which dominated the diuretic/hypertension market until the emergence of ACE (angiotensin-converting enzyme) inhibitors in the 1990s.[16]

Charles Winter at Merck (who later moved to Parke-Davis) developed a test called the cotton string granuloma test as a model of inflammatory pain. Using this model, Merck screened about 350 indole compounds and identified indomethacin as a potent anti-inflammatory drug. Indomethacin was initially synthesized by medicinal chemist T. Y. Shen (who later became vice president of inflammation research at Merck) as a plant growth regulator. It was also found to be particularly active in another model of inflammatory pain, carrageenan-induced rat paw edema. Indomethacin was introduced as Indocin in 1964. In addition, Merck synthesized more than 500 salicylate compounds that led eventually to an anti-inflammatory analgesic, diflunisal (Dolobid), a 5-fluorophenyl salicylate, in 1971.

Bristol-Myers Squibb's ACE inhibitor captopril (trade name Capoten) was introduced in 1978. Merck assembled a group of scientists, led by Arthur A. Patchett, in an attempt to make a better ACE inhibitor by improving upon captopril.[17] Patchett earned his Ph.D. degree from Robert Burns Woodward at Harvard University. He joined Merck after a stint at the National Institutes of Health (NIH). Because of his scientific talents, he was quickly promoted to the head of the entire chemistry operation at Rahway. He was a brilliant chemist, but he was not much of a manager in his early years. Stripped of his managerial post in 1973, Patchett was banished to an old dungeon-like laboratory, making random peptides and doing other menial work. Roy Vagelos, the research chief at the time, often toured the laboratories on Saturdays, and Patchett was always there. During those visits, Patchett had opportunities to talk to Vagelos about chemistry, science, and drug discovery, after which Vagelos promoted him back to leading projects.[18] Under Patchett's leadership Merck arrived at a very potent ACE inhibitor, enalaprilat, which unfortunately suffered from poor oral bioavailability. They simply converted the carboxylic acid functionality into its corresponding ethyl ester, creating enalapril, a prodrug of enalaprilat that had excellent oral bioavailability. Although enalapril is a "me-too" drug of captopril, it is better absorbed by the stomach. One advantage of a prodrug is the delay in onset of action, which can be beneficial for a drug to treat blood pressure. The longer duration of action also allows a once-daily dosage. In addition, enalapril is devoid of the side effects associated with the thiol group present in captopril, including bone marrow growth suppression, skin rash, and loss of taste. In 1981, Merck successfully completed the clinical trials, gained FDA approval, and sold enalapril using the brand name Vasotec, which became their first billion-dollar drug in 1988.

Mevacor and Zocor

Among all the drugs coming out of Merck, none have surpassed Mevacor and Zocor in terms of financial success.

The poor Greek immigrant kid

In 1969, Max Tishler chose Lewis Sarett as his successor as Merck's head of research and development. In 1975, Sarett passed the baton to Roy Vagelos, then professor and chairman of the Department of Biochemistry at Washington University in St. Louis.

Vagelos, born in October 1929, grew up in Rahway, New Jersey, where Merck's headquarters were located. His parents, both poor Greek immigrants, owned a small restaurant called Estelle's Luncheonette in Rahway. Merck was only six blocks away, and many of its research scientists and engineers regularly went to the restaurant for breakfast, lunch, or dinner. While helping his parents with the tables, Vagelos spent a lot of time talking with his customers, whom he got to know very well. What Merck was doing in drug discovery fascinated him, and he decided that he wanted to do work that improved people's lives.

In 1947, Vagelos went to the University of Pennsylvania. Although he loved organic chemistry, he ended up going to medical school at Columbia University four years later. That summer, he worked as an intern at Merck & Co. under Dr. Harry Robinson. His assignment was repetitive and boring, which made medical school even more attractive. In 1955, Vagelos was selected to do his residency at the prestigious Massachusetts General Hospital, where he was trained in dealing with patients firsthand. A year later, he joined the NIH as a research physician, thus fulfilling his military service requirement for the U.S. government.

At the NIH, Vagelos initially worked with Earl Stadtman, an eventual Nobel laureate. There, he also got to know another Stadtman protégé, Michael S. Brown, who along with Joseph L. Goldstein would later discover the LDL receptor in 1973. Guided by Stadtman, Vagelos worked on the biosynthesis of fatty acids. While he was making headway in his independent research, another future important player in the Mevacor saga, Alfred W. Alberts, came to work for him as a research associate. Alberts chose a unique career path. He did his graduate work in cell biology at the University of Maryland. In 1959, he got a job at the NIH working for Vagelos in lipid research, leaving the University of Maryland with an ABD, "all but the dissertation" (meaning that he had completed all the requirements for his Ph.D. but did not have his dissertation written). In 1965, Vagelos was offered a job as the chair of the biochemistry department at Washington University, succeeding Nobel Laureate Carl Cori. Alberts showed great loyalty and followed Vagelos to St. Louis. There, Alberts helped Vagelos run his research group while Vagelos busied himself building a world-class department. Because of his value to the students, the department, and biochemistry in general, Alberts was unanimously recommended to become an associate professor with tenure.

Vagelos said, "When you do good work in hot science, you receive many attractive offers."[18] His pioneering research into fatty acid biosynthesis attracted the attention of scientists from around the world. In the early 1970s, upon the recommendation of Harry Robinson, Merck hired Vagelos as a consultant to help Merck scientists better understand what was going on at the cutting edge of biochemistry and enzymology. One of the cholesterol-lowering drugs that Merck discovered from screening during this time was halofenate, whose mechanism of action was complicated and unclear. Vagelos was horrified by this aspect of unknown mechanism and suggested that there was a better way to lower cholesterol. He gave a lecture on the biosynthesis of cholesterol and recommended that HMG-CoA reductase, the rate-limiting enzyme, should be the target for pharmacologic intervention. At the end, halofenate flopped in clinical trials due to its multiple pharmacologic effects and accompanying toxicity. Impressed by Vagelos, Merck CEO Henry Gadsden recruited him in 1975 as the head of research and development, succeeding Lewis Sarett.

When he joined Merck, Vagelos brought with him several academic friends, including Alberts, who would become the "product champion" of Mevacor. One of the professors that Vagelos "stole" from academia was Christian R. H. Raetz, who had been a professor at the University of Wisconsin. Raetz met Vagelos at a bar at a Gordon Research Conference for Lipid Metabolism at Kimball Academy in New Hampshire in 1971. Vagelos obviously had a good enough impression of him because he hired him as Merck's head of biochemistry.

Vagelos brought to Merck a strong academic culture where science and scientists dominated the company. Being a leader in drug discovery, he practiced what he preached, transforming the way Merck did research. Pretty soon, Merck was one of the pioneering drug firms who first embraced "molecular targets," also known as "rational drug discovery." Vagelos later hired Edward Scolnick from the NIH to replace him as the head of research and development when he was promoted to CEO in 1985.

Prelude

Mevacor is the result of more than 23 years of effort by scientists both inside and outside of Merck. According to Vagelos, "The history of the discovery and development of Mevacor illustrated well the interdependence of basic and applied pharmaceutical research, as well as how long, tortuous, and risky the pharmaceutical discovery and development process can be."[19]

In the early 1950s, Jesse Huff and his associates at Merck began researching the biosynthesis of cholesterol, building on the efforts of scientists outside of the company. In 1956, from a yeast extract, Karl Folkers and Carl H. Hoffman isolated mevalonic acid, a crucial chemical in the series of reactions that produce cholesterol. Huff and his team subsequently demonstrated that mevalonic acid could be converted into cholesterol. One year later, Merck looked for resins that would bind to bile acids as a means of lowering cholesterol. Cholestyramine reduced cholesterol by 10–15%, but the sandlike texture made it unpalatable, and constipation was an unpleasant side effect.

During 1958–1959, the conversion of HMG-CoA into mevalonic acid was shown to be the major rate-limiting step in cholesterol biosynthesis. Building on their own experience with fatty acid biosynthesis, Vagelos and Alberts began to devise a scheme to find an HMG-CoA inhibitor that was an anticholesterol agent. Michael Brown, a consultant for Merck at the time, convinced scientists there in 1976 to begin a screening program, looking for statins from Merck's natural product efforts.

In April 1976, Akira Endo, then at the School of Agriculture at Tokyo Noko University, was invited to Merck in Rahway to give a lecture on the development of his mevastatin. After the seminar, both Vagelos and Alberts discussed the experimental results with him.[20] One month later, Merck's head of research in Japan, H. Boyd Woodruff, and Sankyo's H. Okazaki signed a one-page disclosure agreement with regard to statins. Sankyo granted Merck access to its data and methods connected with Akira Endo's mevastatin. Companies often release such information to potential business partners, but unfortunately this agreement left a gaping hole: Merck did not owe Sankyo anything if it found the same anticholesterol properties in another fungal by-product,[21] which was exactly what happened.

Discovery

In 1974, in an attempt to take advantage of mechanism-based drug discovery, Merck organized the Fermentation Products for Screening (FERPS) project, in which broth extracts were sent for biochemical assays while those assays were being developed for high-volume testing. Meanwhile, Merck hired the small Spanish firm Centro de Investigación Básica de España (CIBE, whose predecessor was Compañía Española de la Penicilina y Antibióticos) in

Figure 3.3 Alfred W. Alberts.

Figure 3.4 Julie S. Chen.

Madrid to run this high-volume screening. Dr. John Rothrock and Ms. Maria Lopez prepared a sample from a culture of *Aspergillus terreus* (a common soil fungus found around the world) and sent it off to Arthur Patchett's group, which made the extracts and forwarded them for various biochemistry assays. One of them was the HMG-CoA reductase assay devised by Alfred Alberts and his laboratory assistant Julie S. Chen. The assay, measuring the formation of mevalonic acid from HMG-CoA, was a rapid, high-throughput enzyme assay that allowed tests to be carried out in large numbers. Chen and Alberts began to screen thousands of soil cultures while searching for HMG-CoA reductase inhibitors.[22]

A breakthrough came on November 16, 1978, just a few days after the HMG-CoA project began. Chen walked into Alberts's office and told him that no mevalonic acid was detected in the assay for the 18th microorganism in the binding experiment using a radioactivity assay—the active principle of the broth potently inhibited the enzyme, HMG-CoA reductase. In fact, the compound was twice as active as mevastatin in rat liver HMG-CoA reductase. This was extraordinary; for most high-throughput assays, it takes thousands of samples to get the first hit. Fearing the data were too good to be true, Alberts asked Chen to repeat it. Twenty-four hours later, the data were confirmed—they had found a winner, a potent HMG-CoA reductase inhibitor in the first week! Backtracking the sample, they identified the microorganism from the culture broth as *Aspergillus terreus*. Ironically, nothing with a better profile was ever discovered from screening additional thousands of samples.

The FERPS project team at Merck led by Carl Hoffman set about isolating the HMG-CoA reductase inhibitor that Alberts and Chen had found. Three months later, in February 1979, Hoffman isolated the pure compound that was later given the generic name lovastatin (trade name Mevacor). Coincidently, 22 years before, in 1957, Hoffman was the one who discovered mevalonic acid, an intermediate in the cholesterol biosynthesis pathway.

The structure of lovastatin was elucidated using spectroscopy by George Albers-Schonberg and his associates in Merck's Department of Biophysics. It differs from mevastatin by merely an additional methyl group. Merck initially tested lovastatin in rats and mice, but they found dogs responded much better. To gain confidence in its safety profile, Merck's pilot plant prepared large amounts of lovastatin and carried out extensive toxicology studies. With confidence in its safety established, Merck quickly moved lovastatin into clinical trials in April 1980. At that point, the lovastatin project team under the leadership of Alberts as the "product champion" had grown from a team of two into a more than 100-person-strong powerhouse.

Endo also independently discovered lovastatin three months *after* Merck in February 1979 at Tokyo Noko University. He subsequently licensed it to Sankyo, which filed for the patent in the same month in Japan and about 30 countries so its patent application came four months *before* Merck's patent on lovastatin in June 1979. As a consequence, Sankyo did not receive patents in the United States and several other countries that gave priority to *time of invention*. However, in some other countries, such as Japan, that give priority to *time of application*, patents were granted to Sankyo rather than Merck. Although Sankyo did commercialize lovastatin and only paid Tokyo Noko University a meager 35 million yen, it would later on receive billions of dollars from licensing by blocking Merck's commercialization of Mevacor in many major countries except the United States.[20]

Halt!

In September 1980, the clinical trials for lovastatin were abruptly halted. Boyd Woodruff, Merck's head of research in Japan, heard rumors that Sankyo's mevastatin development had been discontinued because intestinal tumors in dogs were observed. Since lovastatin and mevastatin differed only by an additional methyl group, they reasoned that if mevastatin was toxic, chances were good that lovastatin was toxic, as well. Worse yet, if the toxicity was mechanism based, all compounds that inhibited HMG-CoA reductase would be toxic! Vagelos made the decision to discontinue the clinical trials of lovastatin. He called Jonathan Tobert,[23] who led the Merck clinical trials for lovastatin, and the trials for lovastatin were immediately brought to a screeching halt. Development of lovastatin was stalled for three years. During that time, Vagelos went to Tokyo to personally meet with the CEO of Sankyo Pharmaceuticals. According to Vagelos, he offered Sankyo a business deal: "If you help us solve this problem, we'll share Mevacor with you in Japan and you can share your second-generation product with us when you are ready."[18] But the head of Sankyo just smiled and replied that he wanted to cooperate, but others objected to any exchange of data. In the end, Vagelos never had a scientific discussion about what Sankyo discovered because Sankyo approached it as a problem of corporate competition rather than a medical problem in drug discovery.

While lovastatin was stuck in limbo, some prominent clinicians who took part in the initial clinical trials started talking to Merck in an attempt to convince them to continue the clinical trials. In early 1982, Drs. Roger Illingworth of Portland, Oregon; Scott Grundy and David Biheimer of Houston, Texas; and Daniel Steinberg of University of California at San Diego urged Merck to restart limited clinical trials for hypercholesterolemia because their patients had dire prognoses. That year, the FDA gave Merck special permission to use lovastatin for the treatment of hypercholesterolemia. Encouraged by the FDA's favorable ruling, Merck reinstituted animal studies in August, and gratifyingly, they did not see any intestinal tumors from lovastatin in dogs. Merck toxicologists carried out elegant studies by dosing lovastatin along with mevalonic acid. Many of the mechanism-based toxicities were prevented by the presence of mevalonic acid. In essence, whatever toxicities were associated with blocking the HMG-CoA enzyme could be reversed with the addition of mevalonate. In November 1983, based on these positive results with the hypercholesterolemia patients, as well as a recommendation by Steinberg at UCSD and Jean Wilson of University of Texas, formal clinical trials of lovastatin recommenced. After negotiation with the FDA, Merck resumed large-scale clinical trials for lovastatin in May 1984. Jonathan Tobert, who led the trials, was ecstatic. In the ensuing two years, the clinical trials and long-term safety studies were completed. At that point, lovastatin took up a quarter of Merck's total research budget of $500 million. These trials decisively demonstrated that lovastatin dramatically lowered plasma cholesterol with very few side effects. No agent had ever exerted such dramatic drops in cholesterol levels. More important, lovastatin was well tolerated, unlike previous cholesterol-lowering agents such as niacin, resins, and fibrates. In October 1986, long-term toxicology studies on dogs were finished and no tumors were noted!

Triumph

On November 14, 1986, Merck sent a van to the FDA loaded with 104 volumes, each containing more than 400 pages, as their package to apply for regulatory approval of lovastatin. Even before the New Drug Application was filed, Merck had many interactions with the FDA concerning the medical benefits and safety of lovastatin. Therefore, the agency officials were already familiar with the details of the compound because Merck had conscientiously kept them informed of the drug's progress every step of the way.

In late February 1987, an FDA advisory panel convened to evaluate various safety issues arising from the animal toxicology studies and clinical trials. Interestingly, the most serious question that the FDA advisory panel raised was whether lovastatin might cause heart attacks rather than prevent them. The panel's concerns were derived from the experience with *dextro*-thyroxine, an early cholesterol-lowering drug that actually caused ischemic heart disease as shown by a large-scale, long-term clinical trial in men. Because lovastatin and *dextro*-thyroxine work through completely unrelated mechanisms of action, the concerns were not justified. In fact, lovastatin's safety was later solidly proven after use by a large population. In August 1987, the FDA approved the marketing of lovastatin in a record time of nine months (vs. the usual 30 months), partially because the FDA was familiar with lovastatin and its history. The FDA stated that, in clinical trials, lovastatin lowered total cholesterol levels by 18–34%, depending on the dosage used. More important, it reduced levels of damaging LDL cholesterol by 19–39% without lowering the protective HDL cholesterol that is thought to cleanse the body of artery-clogging cholesterol. In fact, in most patients taking lovastatin, HDL cholesterol increases by about 10%.[24] In four years of clinical testing, lovastatin was shown to be simpler to take, easier to tolerate, and less likely to cause unpleasant or serious side effects than any other available cholesterol-lowering drug at the time. Michael Brown, who received a Nobel Prize for his lipid research, noted, "With this drug, we now have the ability to control one of the three major risk factors for heart disease: cigarette smoking, high blood pressure and elevated cholesterol."[24]

Merck sold lovastatin under the trade name Mevacor. Interestingly, because of Richardson-Merrell's terrible experiences with its cholesterol-lowering drug MER/29, it was feared that Mevacor might possess the same toxicity in causing cataracts. The FDA therefore advised that the patient for whom lovastatin is prescribed should have blood tests every six weeks to monitor liver function, as well as annual eye examinations. An increase in liver enzymes was been noted in about 1% of patients taking lovastatin, which could mean their livers are being overworked. Other patients experienced changes in the lens of the eye that could suggest an increased risk of cataracts. Before these studies, prestigious medical journals such as *Lancet* and the *New England Journal of Medicine* even published editorials questioning the safety of cholesterol-lowering drugs in general and Mevacor in particular. However, after a few years, because there were no reports of cataract formation in patients taking Mevacor, the FDA eventually removed that requirement.

The isolation of lovastatin was the fruit of four years of Merck's FERPS program. And after testing nearly 5,000 fermentation extracts in numerous enzyme and receptor assays, lovastatin was the *only* active compound isolated! Lovastatin, through the efforts of many individuals, was the first statin on the market. The name Mevacor comes from the root "Meva-" indicating its ability to stop the synthesis of mevalonic acid. The suffix "-cor" was meant to indicate something to do with the heart as in coronary. The trend was followed by several other statin drugs, such as Zocor, Lipitor, and Crestor. The tablet color for Mevacor was initially designed to be yellow. However, Merck scientists felt this color resembled the color of butter, perhaps not a good image for cholesterol reduction. It was later changed to blue.

One drawback to Mevacor, at least from the patient's standpoint, was its initial high price. Three doses were approved by the FDA: 10, 20, and 40 mg. A single 20-mg pill went for $1.64, and a year's treatment could cost up to $3,000. On the other hand, one calculation pointed out that patients taking Mevacor actually saved money in the long run, considering the costs of future surgical and hospitalization expenses if their condition was not treated. Moreover, the profit generated from Mevacor sales afforded the revenue needed for research, testing, and development of new products. Lovastatin has been available in inexpensive generic form since 1999.

For all its potential, Mevacor faced stiff competition. Lopid, a similar drug introduced in 1982 by Parke-Davis, controlled about 40% of the $190 million anticholesterol business when Mevacor appeared on pharmacies' shelves in September that year. But Mevacor quickly grabbed a 33% share, trimming Lopid's lead to 20%. Then, in November, Parke-Davis came out with a study quantifying how Lopid dramatically cut the risk of coronary heart disease. Lacking his own data, Vagelos refused to make similar assertions. By January the two drugs were running neck-in-neck in sales.

In September 1989, author Thomas J. Moore published in the *Atlantic Monthly* "The Cholesterol Myth," an excerpt of his book *Heart Failure: A Searching Report on Modern Medicine at Its Best... and Its Worst.* He cited a "meta-analysis" of statistics collected by epidemiologists. From the methodology the author used, he somehow "proved" that lowering cholesterol corresponded to an increase in mortality![25] Although we know today that Mevacor is extremely safe, at the time the book became a best seller, and sales of Mevacor suffered for some time afterward.

Efforts toward the discovery of additional analogs of lovastatin were undertaken with great success by Robert L. Smith and his colleagues, including Alvin K. Willard and William F. Hoffmann at Merck Research Laboratories in West Point, Pennsylvania. They synthesized simvastatin (trade name Zocor), an analog of lovastatin with an extra methyl group on the side chain. Simvastatin, 2.5 times more potent than lovastatin, is twice as active as mevastatin against rat liver HMG-CoA reductase and longer lasting than lovastatin. More important, Merck held the patent for Zocor all over the world, unlike Mevacor. The FDA approved Zocor for marketing in 1991.

What really established the value of statins was Merck's "4S study"— the Scandinavian Simvastatin Survival Study. The study followed 4,444 patients with coronary heart disease and elevated cholesterol levels (219–310 mg/dL) for a median of 5.4 years. An average patient treated with simvastatin (Zocor) saw a decrease in total cholesterol of 25% and in LDL cholesterol a decrease of 35%, with an 8% increase of HDL cholesterol. This was the first clinical trial to demonstrate conclusively that long-term therapy with Zocor could reduce the recurrence of heart attacks. A 1%

decrease in total cholesterol resulted in a 1.5% reduction in risk of coronary death. Most crucially, the 4S study, for the first time, decidedly demonstrated that lowering cholesterol lowered the rate of both morbidity and mortality. A 1% decrease in total cholesterol resulted in a 1% decrease in total morbidity and mortality. In light of the results of this study, in 1995, the FDA approved Zocor as a product to save lives and to prevent patients who had previously suffered a heart attack from having another. For many years, Zocor was the gold standard for cholesterol-lowering drugs. Zocor also brought a windfall to Merck. From 2000 to 2005, Zocor's sales were $2.93, $5.26, $5.51, $5.00, $5.20, and $4.40 billion, respectively. After Schering-Plough introduced Zetia (ezetimibe), a newer cholesterol absorption inhibitor, Merck and Schering-Plough combined Zocor and Zetia. They have sold the combination drug under the trade name Vytorin since 2004 (more on Zetia and Vytorin later).

When the Swiss drug firm Sandoz Pharmaceuticals introduced fluvastatin (trade name Lescol) in April 1994, the Sandoz management made a conscientious decision to market Lescol with significantly lower prices than other statins. The strategy worked well. In the first year, Lescol earned 60% more prescriptions than Zocor. Countering, Merck reduced the prices of both Mevacor and Zocor by 32%, although the prices of most other drugs rose at the time.

Interestingly, shortly after lovastatin (Mevacor) became popular, it was discovered that lovastatin exists in Chinese fermented red rice. Pharmanex Inc., a small company in Simi Valley, California, took advantage of this observation and began selling Chinese fermented red rice as a natural cholesterol-lowering supplement with the trade name Cholestin. Cholestin was made from a strain of rice fermented with red yeast, which was then ground into a brick-colored powder and pressed into capsules. The rice was imported from China, where it had been used for more than 2,000 years, both as an herbal remedy and a food source. Since such a minute quantity of lovastatin is found in Chinese fermented red rice, the supplement was unlikely to achieve sufficient efficacy as lovastatin did. Nevertheless, prompted by complaints from officials at Merck, the FDA began an investigation of Cholestin and impounded 10 tons of the red rice. In 1998, agency officials declared Cholestin a drug and insisted that the product undergo the same rigorous testing as any pharmaceutical. Pharmanex promptly sued. Judge Dale A. Kimball of the federal district court in Salt Lake City ordered the FDA to permit Pharmanex to continue making Cholestin.

He wrote, "Plaintiff's Cholestin product is preliminarily declared to be a dietary supplement, and not a drug, within the meaning of the Federal Food, Drug and Cosmetic Act."[26]

The Company, the Drugs, and the Inventors Today

The small Spanish laboratory CIBE was initially contracted by Merck to find an inhibitor of another enzyme, dihydrofolate reductase. After lovastatin (Mevacor) was marketed, CIBE received a royalty on its sales until Merck acquired it a few years later. It turned out that it was much cheaper for Merck to buy CIBE than to keep paying it royalties.

Vagelos retired as Merck's CEO in 1994. Under his leadership, Merck became the largest pharmaceutical company in the world. He hired only the best of the best, and working at Merck was every scientist's dream if he or she coveted a job in the drug industry. The Vagelos era was Merck's golden age. What Merck did was the gold standard for the industry and the envy of all drug companies. The 115-year-old Merck & Co. has built a global research organization that ranks among the best worldwide in terms of the caliber of its scientists and groundbreaking medical research. Today, Merck employs about 56,700 employees in 120 countries and 31 factories worldwide. Their products are sold in more than 200 countries.

Alfred Alberts, the scientist without a Ph.D., immortalized himself through his contributions toward the discovery of Mevacor. He retired in 1995 after 20 years at Merck. He was quoted as saying: "Strategies to unearth blockbusters today are not working." He fears that the advantages of today's powerful drug-hunting technologies are offset by what he sees as a loss of freedom to stretch one's mind around the ideas: "Too much computer and not enough brain."[27] Alberts, along with George Albers-Schonberg, Richard L. Monaghan, and the estate of the late Carl Hoffman, were named co-winners of the 1988 Inventor of the Year Award, with a check of $4,000 each.[28] This accolade was given by Intellectual Property Owners Inc., an organization concerned with patents, trademarks, and copyrights. The winners were selected for inventing lovastatin. Robert Smith, the chemist who invented simvastatin (Zocor), is now the Head of Research at Lion Pharmaceuticals.

In 2000, when Merck and Bristol-Myers Squibb applied to have Mevacor and Pravachol approved for over-the-counter (OTC) use, the

FDA turned them down. But in May 2004, the Health Department in Britain made the decision that low-dose lovastatin (Mevacor) could be sold OTC, which means that a patient in the United Kingdom can buy lovastatin without a prescription. Historically, European countries have been more liberal in their transformation of prescription drugs into OTC drugs, whereas the United States has been more conservative. Mevacor lost its patent protection in the United States in 1999. In February 2005, the FDA again rejected Merck's request to sell Mevacor OTC so that a patient would not need a prescription to buy it. The FDA was not really worried about the efficacy and safety of Mevacor, but argued that the patient would be better served under a system that required surveillance from doctor. Meanwhile, a similar request by Bristol-Myers Squibb to sell the 20-mg version of Pravachol OTC was also turned down by the FDA. According to the FDA, the patients did not seem to understand cholesterol well enough to treat themselves. In one study, half the people who thought they were candidates for OTC Mevacor were wrong. Their cholesterol was either so high they needed a doctor's care, or too low to even consider drugs. In December 2007, for the third time, the FDA rejected Merck's request to sell Mevacor OTC. The twist this time is that GlaxoSmithKline had already purchased, for an undisclosed price, the right, which no longer exists, to market the OTC version of Mevacor.

Zocor lost its patent protection in 2003 in the United Kingdom, and the U.K. Department of Health approved statins for OTC sale in May 2004, with Zocor the first to be available at pharmacies. In June 2006, Zocor also lost its patent protection in the United States, and cheap generic simvastatin has been on the market since then.

An episode involving former President Bill Clinton heightened Americans' awareness of coronary heart disease and elevated the status of statin drugs in general, and Zocor in particular.[29] At the time of his exit physical examination as president in 2001, blood tests showed he was at increased risk for heart disease. His total cholesterol was 233 mg/dL, with LDL cholesterol of 177 mg/dL. This was considerably higher than the 137 of the previous year's examination, so White House physicians prescribed Zocor to lower his cholesterol and other lipid levels. However, in September 2004, Clinton underwent successful quadruple coronary bypass surgery at a New York hospital three days after tests prompted by chest pains and shortness of breath. Mr. Clinton, who was 58, took various medications, including aspirin to thin the blood, a statin (Zocor) to lower cholesterol, a beta-blocker to prevent irregular heartbeats, and an ACE inhibitor to control high blood pressure. Mr. Clinton was given Plavix, a blood thinner marketed by Bristol-Myers Squibb, as a precaution to ease the blood flow through his narrowed arteries, and physicians said they wanted to let the drug clear from his system to avoid any excessive bleeding during or after the surgery.[30]

Currently, Merck & Co., along with the pharmaceutical industry in general, is going through tough times. In September 2004, Merck voluntarily withdrew its blockbuster painkiller Vioxx from the market after clinical trials showed that it doubled the risk of heart attack and stroke incidents for patients taking it. In the few years since then, Wall Street has wiped out 38% of Merck's stock value, amounting to $25 billion dollars. It will take much hard work for Merck to reclaim the glorious status it enjoyed before the Vioxx episode.

In 2006, the FDA approved Merck's type 2 diabetes drug Januvia (sitagliptin), a DPP IV (dipeptidyl peptidase-4) inhibitor with blockbuster potential. The same year, Merck launched its revolutionary prophylactic human papillomavirus vaccine (Gardasil) for cervical cancers. Once again, Merck's future was beginning to look bright until early 2008, when a clinical trial showed that Vytorin (the combination of Zocor and Zetia) was no better than simvastatin (Zocor) alone in reducing coronary heart disease.

CHAPTER 4

Discovery of Lipitor

Parke-Davis was not the first drug company to put a statin on the market (Merck was), but it was the one to do it the best, with Lipitor (atorvastatin). When Parke-Davis Pharmaceuticals in Ann Arbor, Michigan, began to look for its own statin in 1982, it was late in the game. Parke-Davis's Lipitor was discovered in the mid- to late 1980s and brought to market in 1997. Already ahead of it were four statins: Merck's Mevacor was released in September 1987 and Zocor in December 1991, Bristol-Myers Squibb's Pravachol in October 1991, and Sandoz's Lescol in March 1994. Despite the competition, by 2006 Lipitor had become the best-selling drug in history, with one-year sales totaling $12.9 billion, more than the net worth of the 10th biggest drug company in the world.

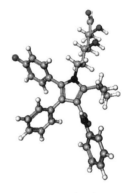

Figure 4.1 Molecular structure of Lipitor.

Splendid History

Parke, Davis & Company

Although the drug firm Parke-Davis had in recent times been relegated to history books, half a century ago, Parke, Davis and Company once enjoyed the status of the largest pharmaceutical manufacturer in the world.

In 1866, 38-year-old Hervey C. Parke, a businessman, and Samuel P. Duffield, a chemist and physician, founded Duffield, Parke & Company—Manufacturing Chemists in Detroit, Michigan. Duffield originally studied under the father of organic chemistry, Justus von Liebig, at the University of Giessen in Germany. He continued to pursue his academic interests even after the founding of the company and published several articles in the *American Journal of Pharmacy*. In 1867, 22-year-old George S. Davis joined the company as the firm's first salesman. Four years later, Parke and Davis bought out Duffield's shares of the company. In November 1871, Parke, Davis & Company was born, with Parke as the president and Davis as the general manager. The company's inventory was typical of the time: aconite, belladonna, ergot, spirit of ammonia, arsenic, and ether. Davis, a Napoleonic, small-statured man, was clearly responsible for building the company's sales and its enterprise in many directions. He pioneered product promotion by publishing books and magazines, a practice later followed by many other companies.[1] When vaccines were first invented in Europe, Parke-Davis was one of the first pharmaceutical companies to move into this new field. By the end of the 1890s, Parke-Davis was the second company to manufacture the diphtheria antitoxin in the United States (H. K. Mulford Co. in Philadelphia was the first).[2] In 1902, Parke-Davis became the first American pharmaceutical company to build its own research laboratory. The first chemist employed at the research laboratory was Thomas Aldrich, who made important contributions to one of Parke-Davis's best-selling drugs: adrenaline.

The hormone isolated from the adrenal glands above the kidneys, adrenaline is the blood-pressure-raising principle of the suprarenal glands. By as early as 1889, it was known that a substance in the adrenal glands exerted a powerful effect on the blood vessels, heart, and muscles. Many renowned chemists, including John Jacob Abel at Johns Hopkins University (who was the first to make crystalline insulin) and Otto von Fürth at the University of Strassburg, tried hard to isolate the active principle without success because the molecule was extremely water soluble and easily oxidized. Japanese chemist Jokichi Takamine, sponsored by Parke-Davis, overcame the oxidation obstacle by carrying out the extraction process under an atmosphere of carbon dioxide. In 1900, Takamine successfully isolated 4 grams of adrenaline as beautiful octahedral crystals by simply adding ammonia to a concentrated aqueous extract of the adrenal glands. Meanwhile, Aldrich, who was Abel's student before joining Parke-Davis, also isolated a small amount of adrenaline in the summer of 1899. Using the combustion analysis technique (burning the compound, weighing the products [water and carbon dioxide] and then backtracking to determine the atomic composition of the compound), Aldrich was the first to correctly determine the precise chemical formula of adrenaline. Two years later, Parke-Davis began commercially producing it. Initially 500 head of cattle were required to isolate just 1 kg of adrenaline. Parke-Davis marketed adrenaline under the trade name Adrenalin,[3] which rapidly found clinical use in relieving respiratory distress, such as asthma attacks and allergic reactions, and to jump-start the hearts of patients who were in cardiac arrest. Soon adrenaline was in every doctor's bag and saved thousands of failing hearts. Adrenaline subsequently brought huge fortunes to both Parke-Davis and Takamine. With his fame and fortune secured, Takamine dedicated his energy to increasing American appreciation of Japanese culture. He arranged for the gift of 3,000 cherry trees to be sent to Washington, D.C., from the Mayor of Tokyo. Not forgetting the company that made his fortune, Takamine also sent hundreds of cherry trees to the City of Detroit, home of Parke, Davis & Company.[4]

Envious of Parke-Davis's monopoly of the adrenaline market, a rival company, H. K. Mulford Co. (later acquired by Sharp & Dohme in 1929, which in turn merged with Merck & Co. in 1953) brought forth a court challenge to invalidate the adrenaline patent. They argued that the hormone existed in nature and that Takamine's work was anticipated by Abel and von Fürth. In 1911, the court ruled in Parke-Davis's favor on the grounds that the process of extraction, isolation, and purification had provided new material that was devoid of some of the disadvantages of the natural product and earlier extracts and therefore could be useful therapeutically. The federal district judge in New York, Learned Hand, remarked with his judicial wisdom: "I cannot stop without calling attention to the extraordinary condition of the law which makes it possible for a man without knowledge of even the rudiments of chemistry to pass upon such questions as these."[4]

Adrenaline was not the only hormone with which Parke-Davis was involved. Dr. Edward Calvin Kendall, a research chemist at Parke-Davis, was the first to isolate the thyroid hormone thyroxine. In 1914, Parke-Davis produced thyroxine commercially by extraction from thyroid glands. After Kendall moved to the Mayo Foundation, he isolated cortisone, a hormone from the adrenal cortex, in 1936. His colleague Philip S. Hench, a rheumatologist, demonstrated that cortisone worked wonders for rheumatoid arthritis. Kendall and Hench shared the 1950 Nobel Prize in Physiology or

Medicine, along with Polish biochemist Tadeus Reichstein, who also isolated cortisone in 1935.

Because of its reputation, Parke-Davis served as a breeding ground for appointments to university faculties and important posts within other companies. American Home Products, Abbott Laboratories, and G. D. Searle & Co. all hired some of their top executives from Parke-Davis around the middle of the twentieth century.

Dilantin

In 1939, Parke-Davis introduced Dilantin (phenytoin) as a treatment for epilepsy, shortly after its efficacy was discovered by Tracy Jackson Putnam and H. Houston Merritt of Boston City Hospital. Historically, the first really effective antiepileptic drug to enter the physician's armamentarium in the 1850s was potassium bromide, which simply worked as a sedative (it was also rumored to be added to the Soviet Army's diet to reduce soldiers' libido). Later on, physicians explored treating epilepsy with other sedatives. They screened barbituric acids in search of the ones that were effective in treating epilepsy but had less sedative side effects. Chief among them, phenobarbital was most successful, although it still possessed some sedative side effects.

In the mid-1930s, Putnam and Merritt initiated a collaboration to discover anticonvulsive drugs. Their approach was quite naive compared with today's sophisticated drug discovery. Trying to improve upon phenobarbital, they combed the Eastman Chemical Company's catalog and other commercial sources for chemicals containing a phenyl group. After that, Putnam wrote to several major drug firms requesting phenyl-containing drugs. None replied except Parke-Davis, which sent them 19 different compounds analogous to phenobarbital.[5] Overall, Putnam and Merritt systematically screened more than 620 compounds using an animal model, with the convulsion introduced by electroshock. What stood out most was Parke-Davis's phenytoin (diphenylhydantoin), which was not only a more potent antiepileptic than phenobarbital, but also was devoid of any sedative effect. Phenytoin was revolutionary because it was the first drug that separated the antiepileptic effect from the sedative action. As a consequence, it had fewer side effects than other sedatives for epilepsy at the time.

After Putnam and Merritt's discovery, Parke-Davis began marketing phenytoin for the treatment of epilepsy under the trade name Dilantin in 1939. Phenytoin was first invented by German chemist Heinrich Biltz in 1908, and Parke-Davis subsequently purchased it in 1909 without knowing its utility. Thirty years later, the patent for phenytoin had long since expired, and Parke-Davis sold Dilantin for pennies. In 1954, a chemical and biologic research team including Loren M. Long, G. M. Chen, and C. A. Miller at Parke-Davis developed Milontin (quazepam), an anticonvulsant for the control of petit mal epilepsy. Three years later, Parke-Davis introduced another epilepsy drug, Celontin (methsuximide). Parke-Davis's suite of epilepsy drugs, Dilantin, Milontin, and Celotin, now covered all of the three major types of epileptic seizure: motor symptoms, sensory symptoms, and mental symptoms. Parke-Davis solidly established its leadership position in the field of epilepsy treatment and gained enormous respect from physicians in the field.

Interestingly, Dilantin became a fixation of Jack Dreyfus, founder of the Dreyfus Fund, a highly successful mutual fund.[6] In 1958, Dreyfus was president of the Dreyfus Fund and a partner of Dreyfus & Co. At 45, while at the absolute height of his fame, he was afflicted with depression. Although the intense part of his depression lasted for about a year, symptoms lingered for more than five years. He saw a neuropsychiatrist six days a week. In 1963, remembering that a little girl was helped by Dilantin for her epilepsy, Dreyfus subsequently asked for a Dilantin prescription to treat his depression. His doctor did not think it would help, but nonetheless prescribed some of the anticonvulsant for him. Taking 100 mg of Dilantin daily for a few days lifted Dreyfus's mood, and all his major symptoms of distress disappeared. His brain was no longer overactive and filled with negative thoughts. Rejuvenated, Dreyfus went to his work full of energy. Intrigued by Dilantin's previously unknown use, he wanted to investigate further. Flush with cash, he financed many clinical trials for this remarkable medicine and unearthed more than 70 different symptoms and disorders, ranging from alcohol and drug addiction to cardiac arrhythmias, that Dilantin could treat. In the late 1960s, he retired from his businesses and worked full time with the Dreyfus Health Foundation to promote Dilantin, the miracle drug. Despite Dreyfus's crusade in advocating Dilantin, the FDA still had approved it only for a single indication, as an anticonvulsant.

The manufacture of Dilantin is quite tricky, which could explain why there are so few generic versions of Dilantin decades after its patent had expired. Dr. Henze of Parke-Davis improved Heinrich Biltz's original synthesis and patented his synthetic process in 1946. It is even likely that a minute amount of impurity might be crucial for its efficacy because pure phenytoin does not work as well as Henze's preparation. Parke-Davis was consistent in manufacturing Dilantin with good quality, and Dilantin dominated the antiepileptic market. In the early 1990s, Parke-Davis's facility in Puerto Rico that manufactured Dilantin ran into trouble with the FDA. Although Dilantin was produced with high quality, the record keeping was spotty and inconsistent. The FDA imposed fines on Parke-Davis for its failure to conform to Good Manufacturing Practice (GMP). The company paid millions in fines, which contributed to a small-scale layoff of about 60 low-level employees.

Chloromycetin

While Dilantin was one of the well-known medicines that Parke-Davis made, what really made Parke-Davis a household name was chloramphenicol (trade name Chloromycetin), the first broad-spectrum antibiotic. Interestingly, the genesis of this drug traces its roots to a moonlight walk by two scientists.

In 1943, two scientists met at a chemical conference on Gibson's Island in Chesapeake Bay, Virginia. They were Paul Rufus Burkholder, an associate professor of botany at Yale, and Oliver Kamm, research director at Parke-Davis. Kamm, a native of Illinois, earned his Ph.D. at the University of Illinois, studying with the legendary Roger Adams. He was hired as a professor by his alma mater, rising to the rank of associate professor before joining Parke-Davis. His textbook *Qualitative Organic Analysis* was widely used.

One evening during the conference, Burkholder and Kamm took a walk under the moonlight. Their conversation turned to antibiotics isolated from soil. At that time, both penicillin and streptomycin were still not well known, but the exploits of René J. Dubos at the Rockefeller Institute for Medical Research were. Dubos, a student of Selman Waksman (who won the Nobel Prize for discovering streptomycin, the first antibiotic effective against tuberculosis), discovered two antibiotics, gramicidin and tyrothricin, from soil. Although they were both potent antibiotics, neither became drugs because of their toxicity. Inspired by Dubos's discoveries, Burkholder and Kamm decided that they would collaborate to look for antibiotics from soil samples as well. After the conference, Burkholder received a $5,000 grant from Parke-Davis for a cooperative antibiotic research program at Yale.[7]

Back at Yale, Burkholder collected soil samples from all over the world to test for antibiotic activity against six types of bacteria. A friend of his, Professor Gerald George Langham at Cornell, was studying the genetics of corn and sesame in Venezuela. From a mulched field in El Valle, Langham collected several soil samples and sent them to Burkholder. Among the more than 7,000 soil samples that Burkholder screened, the one that showed the most promise was culture 65 from Langham. When Kamm visited Yale and saw the spectacular results, he tucked the culture vial in his vest and rushed back to Detroit for more tests. Further scrutiny at Parke-Davis revealed that the sample contained an organism called *Streptomyces venezuelae*. Parke-Davis's John Ehrlich and Quentin Bartz isolated the active principle, chloramphenicol. A chemistry group was assembled at Parke-Davis, consisting of Drs. Harry M. Crooks, Loren Long, John Controulis, and Mildred Rebstock. They elucidated the structure of chloramphenicol, which turned out to be a small, rather simple molecule. Unlike penicillin and streptomycin, whose commercial synthesis was still a challenge, chloramphenicol was relatively straightforward to make. The group quickly came up with a route to synthesize it chemically. Dr. Rebstock, a blue-eyed blonde from Indiana, was named "Woman of the Year" by the National Women's Club.[1]

The first patients to receive chloramphenicol were victims of a typhus epidemic sweeping through Bolivia. In 1947, a Parke-Davis clinician, Eugene Payne, brought a supply of the drug to help treat the infected people. In the General Hospital at La Paz, 22 typhus patients, five of them near death, were given injections of the new antibiotic. All were cured.[7]

By 1949, large amounts of chloramphenicol were being manufactured. Parke-Davis introduced chloramphenicol under the trade name Chloromycetin, which brought a windfall for the company. Sales were $9 million in 1949, and $28 million in 1950.[1] Due to Chloromycetin's remarkable efficacy, it was used liberally for all kinds of bacterial infections. After a while, reports appeared that a few patients with prolonged use developed aplastic anemia, a sometimes fatal disease where the body no longer manufactures red blood cells. Although the incidence was extremely low (about 0.001–0.005%),[8] the families of some victims petitioned to have

Chloromycetin removed from the market. The FDA decided to permit Parke-Davis to continue selling the drug but with a "black box" warning on the package, specifying that it should not be used to treat infections for which other drugs are effective. Sales understandably slowed for a couple of years. But by 1955, Chloromycetin sales soared to $53 million, making it one of the best-selling prescription drugs at that time. Chloromycetin's success undoubtedly propelled Parke-Davis to the status of the largest drug company in the world, overtaking Eli Lilly to capture that laurel.

The pill

Parke-Davis was intimately involved with the discovery and development of the birth control pill. With adrenaline and thyroxine in hand, Parke-Davis strived to maintain its leadership position in the hormone market. In the 1930s, Parke-Davis funded several academicians to continue their work in the hormone field. One of them was Russell E. Marker, a professor at Pennsylvania State College. During his eight years of tenure at Penn State, Marker published 147 papers and secured 70 patents, which were all assigned to his sponsor, Parke-Davis. In 1939, Marker developed a process (the Marker degradation) of transforming sapogenin, an abundant natural steroid, into progesterone. Funded by Parke-Davis, Marker led expeditions in southern Texas and Mexico in the summer of 1940 to collect plants and screen them for their sapogenin content.[9] Two years later, after learning that sapogenins were abundant in certain types of wild Mexican yams, he tried to convince Penn State and Parke-Davis to develop his process but was not successful. Marker simply quit his job in 1943, traveled to Mexico, and collected 10 tons of the yams. He isolated diosgenin from the yams and transformed it into 2 kg of progesterone, more than half of the world supply with a market value of $160,000. Impressed by this feat, a drug company in Mexico City, Laboratorios Hormona, invited Marker to join them in forming a new company, Syntex (from "synthesis in Mexico"), in 1944 to capitalize on his enterprise in progesterone production. In October 1951, Carl Djerassi at Syntex synthesized norethindrone, which would later become the first oral contraceptive pill.

Syntex patented norethindrone in early November 1951. Lacking biologic laboratories, drug development experience, and marketing outlets, they licensed norethindrone to Parke-Davis in order to pursue FDA registration and marketing of the drug in the United States. In the middle of clinical trials, Parke-Davis suddenly chose to exit the contraceptive field for fear of religious backlash and returned the license to Syntex. This cost Syntex two precious years, which was aggravated by the fact that Parke-Davis was unwilling to hand over data to the Ortho Division of Johnson & Johnson, Syntex's new marketing partner.[10] Eventually, Parke-Davis agreed to continue the development of norethindrone with Syntex, receiving FDA approval in 1959.

Meanwhile, the rest of the pharmaceutical industry did not sit still. In August 1953, more than 20 months after Syntex's patent was filed, G. D. Searle in Chicago filed a patent for the synthesis of norethynodrel. Since norethynodrel is a prodrug that is converted to Syntex's norethindrone in the stomach, Djerassi believed that synthesis of a patented compound in the stomach was an infringement of Syntex's valid patent. He pushed Parke-Davis, which by then had backtracked to develop Syntex's norethindrone, to pursue a legal resolution, but Parke-Davis did not concur.[11] In 1957, G. D. Searle was selling a profitable anti–motion-sickness drug, Dramamine, whose active ingredient was Parke-Davis's Benadryl (diphenhydramine). Parke-Davis probably did not want to antagonize a valued business partner.

In addition to the birth control pill, Benadryl was another important Parke-Davis drug. In 1943, George Rieveschl and Wilson Huber at the University of Cincinnati synthesized diphenhydramine and assigned the patent rights to Parke-Davis. In 1946, Parke-Davis began marketing it under the trade name Benadryl, the first antihistamine in the United States for managing nasal congestion and allergies.

Another drug that helped make Parke-Davis famous was PCP (phencyclidine hydrochloride), or angel dust. On March 26, 1956, chemist V. Harold Maddox at Parke-Davis in Holland, Michigan, observed an interesting reaction. When he treated a nitrile with phenylmagnesium bromide, he obtained phencyclidine hydrochloride instead of the expected ketone. It took a while for him to figure out the correct structure, even though the reaction was known as early as 1925. In the late 1950s, Parke-Davis developed PCP as an anesthetic. It produced such extreme reactions during trials that the drug was quickly shelved, although it is now sometimes used legally as an animal tranquilizer. Unfortunately, PCP is now widely abused because it is easily prepared in basement labs. It is fast acting—one feels its effects in a few minutes. The abuser may quickly lapse into coma, hallucinate, or bristle with hostility.

Lopid

Parke-Davis had a long history of lipid research. The discovery and market success of Lopid (gemfibrozil) was a prime example. In the early 1950s, Imperial Chemical Industries Ltd. (ICI) in England discovered that clofibrate possessed high activity for lowering serum cholesterol. ICI had sold clofibrate under the trade name Atromid-S since 1958. Four years later, J. M. Thorp and W. S. Waring of ICI published a paper in *Nature* summarizing clofibrate's effects on lipids.[12] The *Nature* paper incited a flurry of research in the drug industry, resulting in many second- and third-generation fibrates that are more potent and safer than clofibrate.

Parke-Davis's gemfibrozil was the second fibrate, after clofibrate, to hit the market, under the trade name Lopid. The initial sales were mediocre, with $35 million in 1986 and $60 million in 1987.[14] However, Parke-Davis conducted an ambitious, five-year, $40 million clinical study of gemfibrozil—the so-called Helsinki Study examined the blood lipid levels of 4,000 middle-age postal employees in Finland. In this study, 2,000 men were given Lopid, while the other 2,000 received a placebo. The results were published in the *New England Journal of Medicine* in November 1987.[15] Generally speaking, as a rule of thumb, for every 1% drop in total cholesterol levels, there was a 2% drop in the incidence of heart attacks. Surprisingly, in the Helsinki Study, cholesterol levels decreased by about 10% in the group that took Lopid, yet their heart attack rate declined by 34%! The apparent reason, experts said, was that men in the study had also about a 10% increase in their HDL and a 10% decrease in their LDL at the same time as their total cholesterol fell just 10%. It was the first time that Lopid was shown to dramatically reduce heart attacks in men with high levels of cholesterol in their blood. Although Lopid lowered the LDL-cholesterol levels only slightly, it raised the HDL-cholesterol levels in the blood, reducing the risk of coronary heart disease. The new study "was the strongest evidence to date" that such an alteration in cholesterol subtypes could by itself reduce heart disease, said Dr. Basil Rifkind, director of the Lipid Research Clinics of the National Heart, Lung and Blood Institute in Bethesda, Maryland.[16] And Dr. Jussi Huttunen, director of the National Public Health Institute in Helsinki, said, "For the first time we have evidence, conclusive evidence, that we can reduce heart attack rates" by changing cholesterol subtypes.[16]

Because the Helsinki Study proved that the drug actually reduced the risk of heart attacks, the FDA approved Lopid for use in reducing the risk of coronary heart disease in January 1989.[17] This action also extended the patent for Lopid to January 4, 1993, in accordance with a provision of the trade bill passed by Congress in the summer of 1988. The sales of Lopid quickly took off, earning $300 million in 1989, up from $190 million in 1988, making it the most prescribed anticholesterol drug in 1989. Due to gemfibrozil's superior safety profile over clofibrate, it was widely used. Lopid's annual sales reached $371 million in 1991 and $474 million in 1992 before it lost its patent protection.

Warner-Lambert attempted to extend the patent of Lopid by formulating a sustained-release version. If successful, Lopid would have gained a three-year patent extension. But the FDA did not give the green light.[18] Without patent protection, anybody could make and sell gemfibrozil. Ironically, Marion Merrell Dow Inc. received the FDA clearance to manufacture a generic version of gemfibrozil in June 1995.[19] Marion Merrell Dow Inc. was the successor of Richardson-Merrell, which was associated with one of the first failed cholesterol drugs, triparanol (MER/29).

Roger S. Newton

Parke-Davis was a latecomer in the statin field. In 1982, when it set up a team in the atherosclerosis section to look for its own HMG-CoA reductase inhibitors, Merck's Mevacor had already been in clinical trials for two years. At this time, the statin team's head of biology was Richard "Dick" Maxwell, who retired in the spring of 1984. The head of chemistry was Milton L. Hoefle, who retired in 1986. Two younger scientists succeeded to the leadership posts for the project: biologist Roger S. Newton became the project chair, and chemist Bruce D. Roth was appointed the cochair. Under their leadership, the team brought forth Lipitor, which became the biggest drug ever, with annual sales of $12.9 billion in 2007, transforming heart care worldwide. Since these two team leaders played pivotal roles in the discovery and development of Lipitor, the story of Lipitor must begin with them.

Roger S. Newton was born on May 1, 1950, in East Orange County, New Jersey, to an American father and a first-generation Swedish immigrant mother. His father, Holmes, worked for Travelers Insurance in

New York and instilled in Roger a great deal of business sense, which would prove extremely useful when Roger went on to run his own company.

Newton went to Lafayette College in Easton, Pennsylvania, with a double major in biology and French. After graduating with honors in biology in June 1972, he attended the University of Connecticut and earned an M.S. degree in nutritional science two years later. At age 23, Newton had an experience that shaped his career as a drug researcher. Watching his great-aunt wither away from cancer, he developed a fervent desire to help people through research into the cause of human diseases. In the fall of 1974, he went to study for his Ph.D. in nutrition at the University of California in Davis under the tutelage of Professor Richard C. Freedland. His graduate studies at UC–Davis proved to be pivotal to both his professional and his personal life. Part of his duty as a graduate student/teaching assistant was to oversee the nutrition class, where he met a beautiful undergraduate student, Katherine "Coco" Wagner (and no, he did not inflate her grade to gain her favor, he claimed). They fell in love and were married in 1979, and Coco has been his constant support and inspiration since then.[20]

In the next two years, Newton received a postdoctoral grant in medicine and carried out research at the University of California in San Diego under Daniel Steinberg, a well-known expert in lipid research in general and cholesterol in particular. In 1982, Steinberg, along with several well-known clinicians, urged Merck to resume limited clinical trials of lovastatin (Mevacor) for hypercholesterolemia. In August 2007, Steinberg published a book titled *The Cholesterol Wars: The Skeptics vs. the Preponderance of Evidence*, chronicling the miraculous power of the statins to prevent heart attacks and save lives. Back in 1980, in Steinberg's group, Newton studied the effects of hormones and metabolic inhibitors on lipid and lipoprotein synthesis in rat hepatocyte cultures. Extending Michael S. Brown and Joseph L. Goldstein's discovery of the LDL receptors, Steinberg and Newton investigated receptor-mediated degradation of LDL in cultured hepatocytes. When Parke-Davis hired Newton as a senior scientist in August 1981, he was already 31. Before then, his father Holmes would often tease him: "Roger, when are you going to get a real job?" Jokes aside, years of education equipped Newton with the proper training and experience that would prove indispensable to the role he played in the discovery and development of Lipitor.

At Parke-Davis, Newton joined the atherosclerosis section led by Dick Maxwell. Newton helped Maxwell establish a drug discovery program to develop molecules to inhibit HMG-CoA reductase and to increase LDL degradation. In October 1981, he hired his first associate, Catherine S. Sekerke, from Johns Hopkins University. Sekerke recalled that on the first day at Parke-Davis, she spent 12 hours helping Newton prepare rat liver microsomes, from which the HMG-CoA reductase enzyme was isolated.[21] Initially, following Brown and Goldstein's original protocol,[22] they had difficulty isolating enough of the enzyme from chow-fed rats because there was too little in the rat preparations. Because Newton and Sekerke were looking for an antagonist, a larger quantity of the enzyme was necessary. They later discovered that cholestyramine-fed rats afforded greater amounts of the enzyme in the homogenized liver microsomes. Since cholestyramine, a bile acid binding resin, reduced LDL levels, rats compensated by generating more HMG-CoA reductase, thus aiding subsequent isolation. The development of the enzyme assay took the better part of six months, from October 1981 to April 1982. During the development of the assay, they chose Akira Endo's mevastatin as their reference compound because it was in the public domain and the unsubstantiated rumor of in vivo toxicity was not a concern for the enzyme assay. One of the assays they developed was the HMG-CoA reductase assay known as the COR assay. Newton then hired his former assistant at UCSD, Erika Ferguson, who was in the Steinberg group at the time. At Parke-Davis, Ferguson carried out the cholesterol synthesis inhibition assays for the statin program.

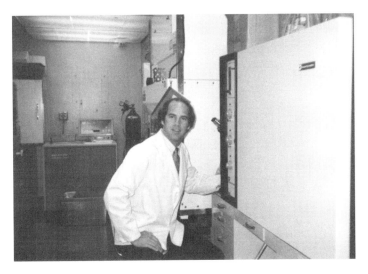

Figure 4.2 Roger S. Newton, circa 1985.

Two years later, Dick Maxwell retired, and Newton was appointed the section head for atherosclerosis biology and chair of the statin team. Under Newton, the biologists included, in addition to Sekerke and Ferguson, Tom Bocan, Brian Krause, Sandy Bak-Mueller, Karen Kieft, Richard "Dick" Stanfield, and Paul Uhlendorf. Together, they established the initial screening strategy for identifying HMG-CoA reductase inhibitors and developed the binding assay using the HMG-CoA reductase enzyme and animal models. The daily enzyme inhibition assays were carried out by Sekerke and Ferguson. The in vivo studies, carried out in the laboratories of Bocan and Krause, were performed using Sprague-Dawley rats after experiments with several species of rats showed these were the most effective. With the assays and animal models in place, all the team needed to be successful was a good small-molecule inhibitor from their chemistry colleagues. It sounds easy, but in reality, despite their ingenuity and hard work, very few medicinal chemists have experienced the thrill of seeing their work directly resulting in a marketed drug. Fewer still help invent the best-selling drug on the planet, unless, of course, you are Bruce Roth.

Bruce D. Roth

Bruce Roth, born on February 16, 1954, grew up in Media, Pennsylvania, just south of Philadelphia. His parents both worked for ACME, a local grocery store, where he also worked 20 hours a week during both high school and college. Although his initial professional ambition was astronomy, a high school chemistry teacher steered his career aspirations toward chemistry. In 1972, he went to St. Joseph's College, a Jesuit school in Philadelphia. Roth, the first in his family to graduate from college, obtained his B.S. degree in chemistry in 1976. During his college years, Roth carried out undergraduate research with Professor George Nelson. Nelson, a Minnesotan, encouraged him to pursue graduate studies in a school in the Midwest, "where one could get a decent education." Under Nelson's influence, Roth moved to Iowa State University in Ames, working for a new professor by the name of George A. Kraus. Before Roth's departure from Philadelphia, the manager at ACME tried very hard to convince him to stay on with the grocery store because of his potential to be a store manager, but Roth chose graduate school instead.

Kraus, Roth's Ph.D. adviser at Iowa State, had recently finished his Ph.D. with Gilbert Stork at Columbia University. As an assistant professor driven to earn tenure and make a name in the field of organic chemistry, Kraus worked hard and expected the same from his students. He stationed his office in the laboratory and worked shoulder to shoulder with his students—"Kraus's hard chargers." Roth worked just as hard in the laboratory and published seven papers ranging from synthetic methodology to natural product synthesis in his five years at Iowa State. Normally, one or two publications would suffice for a Ph.D. degree.

After earning his Ph.D. in 1981, Roth took a postdoctoral fellowship position in the laboratories of Andrew S. Kende, Kraus's mentor for his undergraduate research, at the University of Rochester in New York. Half a year into his stay in Rochester, a neighbor of Roth's ran into him on a Saturday night, a rare occasion, and commented: "You must live a wild life because you are never home!"[23] A wild life indeed he was living—in the laboratory! In just more than one year in Rochester, Roth published with Kende three articles in prestigious journals such as *Tetrahedron* and *Journal of the American Chemical Society*. His experience in the Kende group would later prove very important in the discovery of Lipitor for his experience in working with pyrroles. In 1982, at age 28, when he was offered a job at Parke-Davis as a "scientist," he knew little about Parke-Davis and its parent company, Warner-Lambert.[23] The company did not bestow on him the title of "senior scientist" (as it had done with Newton) because he had not completed the usual two-year stint like most postdoctoral fellows hired by Parke-Davis at the time.

In April 1982, Roth started at Parke-Davis, where he met Michelle Piasecki, a colleague in the PDM (pharmacokinetics and drug metabolism) department, who had started just a week before him. Having not shaken off his habit of working on the weekends, Roth often spent his Saturdays in the laboratory. One Saturday, Piasecki happened to meet Roth on a visit to the laboratory and began talking. One thing led to another, and they were married in 1984.

Three years later, the whole nation watched in horror as the space shuttle *Challenger* exploded on January 28, 1987, killing all seven astronauts on board. The tragedy triggered a life-changing event for Roth. A few months later, lying in bed thinking of the astronauts and pondering the meaning of life, he suddenly began to sweat profusely. He was in such a panic that his wife thought there must have been an earthquake. Soon afterward, he asked his wife Michelle why she was not afraid to die; she shared her faith with him and together they began to attend a church in their local community. Roth was transformed from an atheist to a born-again Christian, giving up all his vices (drinking, smoking, and swearing) as a testimony to his new-found belief. Coincidentally, a year later, their first child, David, was born, although they had tried to have children without success before then. Two daughters and another son would soon follow.

In 1986 the chemistry lead for the statin project and director of the atherosclerosis section, Milton Hoefle, retired. Upon Hoefle's recommendation, the senior vice president of chemistry, John Topliss, of the Topliss Tree (a decision tree for medicinal chemists) fame, appointed Roth as the project cochair for the HMG-CoA project. By then, the statin team in the atherosclerosis section had grown to 18 members. The chemists on the team were C. John Blankley, Alex Chucholowski, Paul Creger (the inventor of Lopid), Mark Creswell, Ann Holmes, Pat O'Brien, Joe Picard, W. Howard Roark, Drago Robert Sliskovic, Charlotte Stratton, and Michael W. Wilson.

Before Lipitor's launch in February 1997, it had to overcome numerous scientific, strategic, and logistic hurdles. The scary part was that any single one of these roadblocks could have derailed the drug. "The number of factors, internal and external, that had to come together for the drug to be a success really boggles the mind," reflected Roth.[24]

The Discovery of Lipitor

Shoulders of giants

Scientific discoveries do not occur in a historical vacuum, nor do they occur in an intellectual vacuum. The seventeenth-century scientist/inventor Isaac Newton (genealogic connection to Roger Newton is unknown) commented that the reason why he could see so far was because he stood on the shoulders of giants. The same can be said of modern drug discovery, including the discovery of Lipitor.

When the Newton-Roth team began their quest for novel statins, they decided to synthesize their own molecules rather than fishing them out of microbial cultures like Merck did. There are two types of statins, natural statins and synthetic statins. The first three statins to reach the market,

Mevacor, Zocor, and Pravachol, were all natural products isolated from fermentation broths, although strictly speaking, Zocor is a semisynthetic statin. In 1981, Merck's Alvin K. Willard, Gerald E. Stokker, and their colleagues filed a couple of patents[25] on their synthetic versions of Mevacor (lovastatin). In place of hexahydronaphthalene as the core structure, they substituted a diphenyl group with four substituents. This demonstrated that it was not necessary to have fermentation products to have the activity against HMG-CoA reductase. Willard's diphenyl-based compounds were active both in vitro (in the test tube) and in vivo (in the body), indicating that synthetic molecules could replace the natural statins. In essence, humans could mimic in the laboratory what the microbes did in making statins.

Sandoz Pharmaceutical Corporation (now part of Novartis) was the first to introduce a synthetic statin to the market. Their fluvastatin sodium (trade name Lescol) was truly innovative. Lescol, the fourth statin on the U.S. market, was an example of a fully synthetic molecule that was just as effective as the natural counterpart. It was invented by Faizulla G. Kathawala at Sandoz in New Jersey in 1984. Kathawala, known as Faizu to his friends, obtained his M.S. from the University of Bombay in India. He earned his Ph.D. in synthetic organic chemistry in 1961 at Technische Hochschule Braunschweig, West Germany, under Professor H. H. Inhoffen. He probably held a record for the number of postdoctoral fellowships at the time, apprenticing under chemistry genius Robert Burns Woodward at Harvard University, H. Muxfeldt at the University of Wisconsin, and Friedrich Cramer at Göttingen University, a school made famous by the father of organic chemistry, Justus von Liebig. Kathawala began working for the Sandoz Research Institute in East Hanover, New Jersey, in 1969. Sandoz was one of the three giant Swiss drug firms, along with Ciba-Geigy and Roche. Not only did he make significant contributions to Sandoz's drug discovery programs, but he was also a very prolific author, publishing numerous review articles and book chapters to serve the chemistry community. By the time Sandoz started its own statin project, Kathawala had risen to director of atherosclerosis/lipoprotein metabolism in charge of the scientific direction of a large group of scientists.

Natural statins such as Mevacor, Zocor, and Pravachol all possess the hexahydronaphthalene core structure. Initially, Kathawala and his colleagues wanted to synthesize the hexahydronaphthalene structure similar to that of mevastatin. However, this core structure has eight chiral centers, and if one were to make it randomly, as many as 256 isomers could be generated! In the end, the chemistry was so challenging that Kathawala decided to give up.[27] Running out of options, Kathawala questioned the importance and the necessity of the complex stereochemistry and substituent pattern present in mevastatin's hexahydronaphthalene core. He then shifted his attention to identify replacements for the core. Naphthalene was his first logical alternative, followed by indole. To the indole core, Kathawala attached a side chain resembling those found in natural statins. The resulting compound would later become fluvastatin sodium (Lescol). Unlike mevastatin, Mevacor, Zocor, and Lescol's side chain was a sodium carboxylate salt instead of a lactone ring. The lactone is actually a prodrug because it is converted to the linear ring-opened form *in vivo*. In addition, Lescol was more potent than both mevastatin and Mevacor. Depending on the particular assay, Lescol was 22- to 146-fold more potent than mevastatin and 10- to 52-fold more potent than Mevacor. At the time, asymmetric synthesis of the side chain seemed unattainable, so Sandoz decided to make fluvastatin sodium in racemic form (a mixture of two isomers, one active and the other less active), which was not as potent as the optically pure drug—all natural statins are optically pure because they consist of only the active isomer. Fortunately, the drug was so much more potent than Mevacor that Sandoz saw no need to increase the potency by another two-fold by separating the two enantiomers. Because of Kathawala's major intellectual input toward the genesis of fluvastatin, he was listed as the sole inventor of fluvastatin in the patents.[26]

In addition to being a mixture of two enantiomers, fluvastatin had another drawback: the presence of a double bond between the indole core structure and one of the two alcohol groups. This type of alcohol (an allylic alcohol) does not survive well in the acidic environment within one's stomach. Thus, it is possible that Lescol would decompose to a certain extent in the gut, which may explain its low efficacy in vivo. Interestingly enough, a latecomer, Crestor by AstraZeneca, also possesses the same side chain as that of Lescol with an allylic alcohol. However, Crestor's allylic alcohol is attached not to an indole but to pyrimidine, which deactivates the allylic alcohol, making it more stable. The indole in Lescol actually activates the allylic alcohol on the side chain, making it more reactive in the gut.

Roger Newton jokingly called Lescol "less cholesterol lowering," although Sandoz initially derived "Lescol" from "less cholesterol." While Lipitor and Zocor boasted billions of dollars in sales, Lescol never reached blockbuster status by reaching $1 billion sales level. Because Lescol was the fourth statin on the market, Sandoz intentionally priced it 50% lower than other statins—Mevacor, Zocor, and Pravachol cost $70 to $80 for a month's supply, but Lescol was sold for only $40 when it was launched in March 1994. Lescol's annual sales reached its peak at $734 million in 2003.

The inventor of Lescol, Kathawala, has since retired from Sandoz in 1994, although he remains quite active in the drug discovery scene, sitting on scientific advisory boards and serving as a scientific consultant to other drug companies. In one of his review articles, Kathawala vividly described the state of competition among all drug firms in the mid-1980s:

> Since the appearance of Merck & Co., Inc. and Sandoz patents and publications, extensive efforts have followed in many laboratories worldwide with semi-synthetic and total synthetic HMG-CoA reductase inhibitors.... It is no wonder that in such a feverish pursuit of finding patentable HMG-CoA reductase inhibitors, review of patent and published literature presents overlapping activities in the laboratories of competing pharmaceutical research companies.[27]

Several other companies, including Rohm & Haas, Hoechst, Rorer, and Bristol-Myers Squibb, were all working on statin projects at the same time. An interesting phenomenon took place among the scientists around the world who were working on statins: they were of course competitors, yet each accumulated knowledge that would teach others and fuel others' own creativity and innovation. The success of statins that benefited millions was the success of teamwork, not only within individual project teams, but also among scientists from academia, industry, and the government. Each contributed one way or another to the statin enterprise. The way by which statins were discovered should serve as the model for modern drug discovery.

Endo's discovery of mevastatin represented a quantum leap of discovery. However, rarely does the first discovered drug possess all the ideal attributes for that class of drugs. In the case of statins, it took incremental improvements by the drug industry as a whole to arrive at the "best in class" drug—Lipitor. This paradigm worked well for the last century, with examples including penicillins, histamine receptor antagonists, ACE inhibitors, and calcium-channel blockers.

The first success… and disappointment

Back to the mid-1980s, the Parke-Davis team was not prescient enough to know which replacement of the hexahydronaphthalene core structure would be most ideal. They chose mevastatin and Willard's synthetic diphenyl statin as benchmarks and began the journey to find their own statins. The chemistry team decided to replace Willard's diphenyl structure with various heterocycles. Such replacements are known as *bioisosteres* because they could potentially exhibit biologic properties similar to the original chemical structure.

In order to substitute Willard's diphenyl structure, Roth recalled a methodology for pyrrole synthesis that he used during his postdoctoral training at the University of Rochester. When later asked why pyrrole was chosen as the template, Roth replied: "There wasn't any 'rational' reason for selection of pyrrole except that I knew how to make it. As a synthetic chemist, often it is that knowing how to make it to help you select the target. In addition, you want to have the ability to incorporate a wide variety of substituents, which was the case for pyrrole."[23] In retrospect, pyrrole might not have been the first core structure Roth would have tried if he had known more history of drug discovery. Pyrrole is an electron-rich heterocycle, which has a tendency to react with electrophiles in the body, causing toxicities. Upon hearing that Roth was working on pyrroles, John Topliss, head of the chemistry department, pulled him aside and commented: "Bruce, you should know that pyrroles do not really make good drugs."[23] Indeed, nowadays, medicinal chemists look at the pyrrole structure with trepidation and avoid it if they can. Being a little stubborn, Roth still pushed it ahead. The team first made dozens of 1,2,5-trisubstituted pyrroles with fluorophenylpyrrole approximating Willard's diphenyl core. In the early 1980s, asymmetric synthesis was still in an embryonic stage, and the technology for controlling chirality was not readily available. Therefore, all the pyrrole compounds were made in racemic form. Within a short time, the team prepared approximately thirty 1,2,5-trisubstituted pyrroles. Unfortunately, even the most potent compound that they made in this series was 10-fold *less* potent than mevastatin in the cholesterol synthesis inhibitor assay using rat liver homogenate.[28] Therefore, the project was still quite far from where they wanted it to be. An order of magnitude increase in potency was needed.

Just when it seemed that the pyrrole idea was stymied by lack of potency, Roth superimposed his pyrrole compound onto Willard's diphenyl statin. This seemingly simple exercise afforded a great insight: Roth's fluorophenylpyrrole was too small to fill most of the space of Willard's substituted fluorodiphenyl core. There was a large region of Willard's compound that was not occupied by the pyrrole compound. In order to fill this region so that the molecule would bind the reductase more tightly, Roth prepared the 3,4-dichloropyrrole analog PD-120167, which was as potent as mevastatin. Shortly thereafter, Mike Wilson, Roth's associate, synthesized the 3,4-dibromopyrrole analog PD-121149, which was also as potent as mevastatin. Now the potency was where the team desired it to be. Wilson, a native of Indiana, had just joined the company after obtaining his M.S. from Purdue University. Although Roth's dichloro analog and Wilson's dibromo analog were equally potent in vitro, Wilson's dibromo analog was more potent in vivo for reducing cholesterol in animals. After pharmacokinetics studies, the team chose the sodium salt of PD-121149 as their lead compound because it was very bioavailable and very efficacious. They promoted this lead compound to clinical investigation status and designated it CI-957. That was the first real success for the team.

No sooner had the team finished drinking their celebratory champagne than Parke-Davis's toxicology department discovered that those halogenated pyrroles, including CI-957, were subacutely toxic in Wistar rats.[29] In fact, at high dose of 600 mg/kg, CI-957 was lethal to rats: the team had come up with the most potent rat toxin ever discovered! It turned out that such exaggerated pharmacology at high doses was not unique to CI-957—this happened to most, if not all, HMG-CoA inhibitors that achieved high plasma and tissue concentrations. The compound was promptly dropped from development like a hot potato. Years later when I spoke with Mike Wilson, he mentioned that CI-957 was actually not a very stable compound. During its synthesis, the bromine atoms sometimes fell off, and he had to reinstall them at the end of the synthesis. Indeed, using today's standard, it probably would never have advanced to the development stage. But back in the mid-1980s, drug discovery was not as sophisticated as it is now. There were very few in vitro toxicology assays to gauge a compound's safety profile. Trial and error was still the primary pathway: you gave the drug to animals and observed the toxicology in escalating doses. If the drug were discovered today, chances are the toxicity would have been detected much earlier using a battery of in vitro assays.

The second refrain

In drug discovery, analogous to great musical compositions, sometimes minor variations of established themes can have a tremendous impact on the success of the work. Listening to symphonies by the great masters, one wonders why they have withstood the test of time and remain popular regardless of how the world has changed. Take Tchaikovsky's Piano Concerto No. 1 (in B-Flat minor, opus 23) as an example. The main theme is replayed seven times, each with different modulation and emotion. In the end, the combination of theme and refrain come to an ultimate and perfect summary, ending the concerto with a resounding note. Like Tchaikovsky's concerto, the second refrain in Parke-Davis's quest of statins came in mid-1985, when Bob Sliskovic synthesized PD-123588, a pyrazole analogue. Chemically, the core of Sliskovic's pyrazole compound differed from Roth's pyrrole by only one additional nitrogen. PD-123588 was as potent as mevastatin in a cholesterol synthesis inhibitor assay using rat liver homogenate.

Drago Robert Sliskovic was born in England to a Croatian father and a British mother.[30] He earned his Ph.D. in organic chemistry from the University of Keele in 1982 under the tutelage of Professor Gurnos Jones. Like many of his compatriots, he came to America to carry out his postdoctoral training. He apprenticed with Frank Lin at the Lederle Laboratories of the American Cyanamid Company in Pearl River, New York. In 1984, he was hired by Parke-Davis as a senior scientist, with Bruce Roth as his supervisor, working on the HMG-CoA reductase project.

Initially, along with Roth, Wilson and Sliskovic worked on Roth's idea on pyrroles. After CI-957 failed, Sliskovic went off to replace the pyrrole with the pyrazole. Overall, the team took a "shotgun" approach, exploring a variety of heterocycles in place of Roth's pyrrole. Eventually, each chemist on the team took on a different heterocycle as the core structure. Roth chose to continue with his pyrrole; Sliskovic chose pyrazole; Chucholowski, pyrimidine; Picard, quinoline; and Creswell, pyridine.

After months of work in assembling the core pyrazole structure and attaching the best side chain similar to that of mevastatin, in mid-1985, Sliskovic finally arrived at PD-123588, the very first compound prepared in this series.[31] It was as potent as mevastatin and possessed the necessary pharmacologic properties. Meanwhile, Roth replaced the two halogens on his pyrrole core with phenyl and phenyl amide. By now, all five available positions of the pyrrole core were fully substituted, with the same side chain as that of Mevacor in the 1-position. In June 1985, Roth, at the tender age of 31, synthesized a racemic lactone that he registered as PD-123832, which would in time lead to the discovery of Lipitor.

At that time, Sliskovic's PD-123588 and Roth's PD-123832 were the two front-runners to be moved into preclinical toxicology studies. Their compounds had very similar potency and efficacy in vivo—both had almost the same efficacy in the EH hypercholesterolemia rabbit model. The atherosclerosis team decided to push Sliskovic's PD-123588 toward development because it was of a chemical series different from Roth's pyrroles. The decision to shy away from Roth's pyrrole series was undoubtedly influenced by the toxicity observed for CI-957. As a result, Sliskovic's PD-123588 became a lead compound in November 1985.

In order to further develop PD-123588, the team needed large enough quantities of the active pharmaceutical ingredient. That meant that they first had to know how to make it. At that time, it was the originating chemists' responsibility to scale up the synthesis to make 100 grams of the compound. But the chemistry that was originally devised by Sliskovic was not amenable to large-scale production, so the team spent several months modifying the synthesis so they could make enough of the compound for preclinical toxicology studies. Beginning in November 1985, Sliskovic and Wilson prepared the 100 grams of PD-123588 required for early development. The synthesis involved nine steps, and at least two steps gave multiple products.[31] Meanwhile, the competition did not stand still. Development of HMG-CoA reductase inhibitors was an extremely competitive field, and many drug firms were working on it. In the spring of 1986, while the Parke-Davis team was still working on their 100-gram lot of PD-123588, they learned that the compound was covered by a patent previously issued to a chemist named James R. Wareing of Sandoz Pharmaceuticals in Switzerland.[32] This patent would prevent others, including Parke-Davis, from using it. Fearing that licensing the patent from Sandoz would be highly unlikely (Sandoz historically had done few licensing deals), Parke-Davis decided to abandon the pyrazole series. This event was a crushing defeat for the team, and they had to regroup and examine their remaining options. In drug discovery, one needs to have persistence—few, if any, compounds a chemist makes will become marketed drugs.

CI-971

After the team decided to drop PD-123588, the only remaining option seemed to be Roth's PD-123832. PD-123832 was a more complicated molecule than the dihalogen compounds PD-120167 and PD-121149. Starting with these dihalogen compounds, Roth first sought to remove the "offending characteristics," namely, the two halogen atoms. Unfortunately, the old chemistry did not work to install the requisite substitutions. After retooling the chemistry for the synthesis of fully substituted pyrroles, Roth substituted the two halogens with a phenyl and a phenyl amide, respectively. By now, all five available positions of the pyrrole core were fully substituted, with the 1-position bearing the same side chain as that of Mevacor. Due to the difficulty in synthesis, a total of only 20 analogs were prepared. In June 1985, Roth first synthesized PD-123832.[33] It was in racemic form, containing both the more active and the less active form. Little did he know then that the more active half would become Lipitor years later. Despite the presence of 50% of the less active compound, PD-123832 was almost as potent as Mevacor.

PD-123832 passed the first hurdle of in vitro pharmacology after Newton's team confirmed its potency in both binding and cholesterol synthesis inhibition assays. PD-123832 lowered cholesterol in the EH rabbit model. Although not significantly more potent in animals than existing statins, PD-123832 did overcome the in vivo pharmacology hurdle that the team needed in order to justify advancing further with this promising new chemical. Newton's team gave it to both casein-fed EH rabbits and cholestyramine-primed dogs with hypercholesterolemia to gauge if it worked in lowering cholesterol. Indeed, the compound lowered serum cholesterol with potency and efficacy similar to that of Mevacor. In September 1986, after the efficacy studies, PD-123832 was brought up for development and designated as a lead compound. The compound was later given clinical investigation number CI-971 in April 1987.

But for a drug to be successful, the next hurdle is bioavailability—how well the drug survives the stomach, crosses intestinal wall, and is absorbed into the bloodstream. No matter how potent a drug may be, if it is rapidly metabolized into inactive metabolites, it will not make a good drug. Back in 1985, drug discovery and development were handled quite differently from now. A drug candidate is now tested for its

PK/PD (pharmacokinetics and pharmacodynamics) very early in the development process, and the ones that do not pass the hurdles are weeded out early. But at that time in the mid-1980s, PK/PD was not done until one had a lead compound such as CI-971. The experiment on bioavailability of CI-971 was performed in Parke-Davis's PDM department, where Roth's wife Michelle worked (and no, they did not give her husband's compound inflated bioavailability). When CI-971 was given to dogs in gel capsules, the bioavailability was only 3%, although the bioavailability was at an acceptable level (greater than 35%) when it was given in solution. The drug was very difficult to synthesize; losing 97% when given to patients was clearly unacceptable. There were two solutions: either reformulate the drug to increase its bioavailability or make the open-chain analog as the dihydroxy acid salt. The team pursued both approaches, but the latter one would prove more successful in boosting bioavailability to an acceptable level. Because the open-chain analog preparation and reformulation would take about six months, the team also decided to develop a chiral synthesis in the interim.

After receiving the green light from both pharmacology and PDM, the next step was to assess the drug's safety. The majority of drugs in discovery phase fail because of toxicity; another large portion of drugs fail for lack of efficacy. Parke-Davis's halogenated pyrroles were discontinued because of their subacute toxicity in rats. This time, Parke-Davis's Department of Toxicology gave CI-971 a clean bill of health for the efficacious doses.

At the time, Newton was personally disenchanted with the support and resources that the Parke-Davis statin project was receiving. In his way of thinking, with only 12–15 people, they were no competition for Merck, which had more than 100 scientists on its statin project. His colleagues would joke with him in the corridor: "Roger, good luck with your project, I heard that Merck now has thirty chemists on their HMG-CoA project."[20] He felt there was no hope of winning the battle against that heavyweight. Discouraged, he almost left Parke-Davis for a position at Bristol-Myers as the head of their cardiovascular biology. Poised to leave Parke-Davis, he made a last effort, pleading with the management for more resources. Ronnie Creswell, Parke-Davis CEO, was convinced enough to double the size of the team to about 30–35 scientists. His support won Newton back to stay. Years later, Newton joked that it was probably a good decision for him to remain at Parke-Davis. In that same year, Bristol-Myers merged with Squibb to form Bristol-Myers Squibb; the position for which Newton had interviewed was eliminated after the merger.

CI-981

Because the statin field was intensely competitive, the Parke-Davis team was very conscientious in finding advantages of their CI-971 over other statins already on the market. But CI-971 behaved similarly to other statins in terms of its efficacy and safety profile.[23]

In addition to not standing out among other marketed drugs, Roth's CI-971 was a racemate, containing two enantiomers. The two enantiomers were biologically distinct: one half was the more active compound and the other half was a less active compound. Initially, the decision was made to bring the racemate CI-971 forward simply because in the mid-1980s, chiral synthesis was still in its infancy and most drugs were developed as racemates if they were not derived from fermentation or from other natural sources. However, it would certainly be better for the patient if the less active molecule was removed, so the liver was not burdened by metabolizing something the patient did not need. Sliskovic was assigned the task of separating the racemic CI-971. Because Willard at Merck carried out similar separations of their synthetic diphenyl statins, Sliskovic simply applied Merck's procedures to CI-971. At first, he treated CI-971 with an optically pure amine, R-(+)-methylbenzylamine. The reaction gave two diastereomers as two amine salts of both the left-hand acid and the right-hand acid. Diastereomers, unlike enantiomers, are easier to separate, which was exactly what Sliskovic did. After separation of the two diastereomers, hydrolysis with sodium hydroxide in refluxing ethanol then provided both the left-hand acid and the right-hand compounds in optically pure forms. With help from Tim Hurley, an analytical chemist in the department, Sliskovic determined that the active right-hand molecule had a (+)-optical rotation and the less active left-hand molecule had a (–)-optical rotation. Later on, the optically pure (+)-compound containing an open chain acid became PD-134298.

Since the open-chain PD-134298 is an acid, making it into a salt would improve both its solubility and physical appearance. As a matter of fact, salt selection is an integral part of drug development. Historically, for acidic drugs, most of the FDA-approved, commercially marketed salts—more than 60%—are sodium salts. Among the statins, Novartis's Lescol and Bristol-Myers Squibb's Pravachol are both sodium salts, as was Bayer's Baycol, now off the market. Potassium salts are the second most frequent for marketed drugs—about 10% of the acidic drugs on the market are potassium salts, and another 10% are calcium salts. AstraZeneca's Crestor is a calcium salt. Zinc, lithium, magnesium, and aluminum salts are all less than 3% in marketed acidic drugs. Overall, for all marketed drugs, there are about 50 types of salts that are sanctioned by the FDA.

When Roth prepared the sodium salt of CI-971, unfortunately, it was hygroscopic, absorbing water when exposed to air and becoming a gooey material. The team had to abandon Roth's sodium salt. Potassium salt was also hygroscopic. While Roth and O'Brien worked on monovalent salts, Sliskovic was assigned the task of preparing bivalent salts, including calcium, zinc, and magnesium. After Sliskovic prepared the calcium salt, it was determined to possess better physical properties—it was registered as PD-124488-38A. The suffix -38A signified that it was the *hemi-calcium salt*. As the project progressed, PD-134298-38A, the hemi-calcium salt of the more active enantiomer, became the lead compound.

In studies performed under carefully controlled conditions with animals, PD-134298-38A was as potent and efficacious as Mevacor in lowering LDL cholesterol. Because it was a salt of the dihydroxy acid, the solubility and bioavailability were considerably improved. The real bioavailability was higher than the amount of free drug available in blood serum because many of the metabolites were themselves potent and efficacious HMG-CoA reductase inhibitors. Even today, there is still debate regarding how much the active metabolites contribute to the drug's efficacy. After breezing through PDM and safety, PD-134298-38A was designated as CI-981, ready for development in clinical trials. It was later christened with the generic name atorvastatin calcium.

With CI-981 in hand, one may think that it would sail through clinical trials smoothly. The reality was far from it—if finding a good drug in discovery was hard, it was even harder for it to reach the market. Indeed, as we shall see in chapter 5, there were numerous occasions when CI-981/atorvastatin calcium could have failed, coming to rest in the graveyard of drugs that never made it to the market.

As for the inventor of Lipitor, Roth steadily rose in the ranks at Parke-Davis. He was promoted to section director of atherosclerosis in 1990, director of atherosclerosis and exploratory chemistry in 1992, and senior director of atherosclerosis and inflammation chemistry in 1993. In 2000, after Pfizer purchased Warner-Lambert, he was appointed vice president of chemistry in Pfizer Global Research and Development at the Ann Arbor laboratories, in charge of a department of more than 200 chemists. In the early 1990s, when Roth was initially promoted to a manager, he tried to balance some laboratory work with his managerial responsibilities. One day he set up a reaction and went to his office to make some phone calls to the biologists. In the middle of his conversation, the acetone cooling bath was ignited and a fire promptly started. That day, his labmates made him realize that he had become a liability to them, literally kicking him out of the laboratory.

With increasing power came increasing responsibilities. Roth began to focus on developing junior colleagues and creating a suitable environment for them to succeed. His boss, Jim Bristol, praised him as a "natural leader." His department became very productive and well respected. Although no longer working in the laboratory, Roth became a manager who "studied failure." In an interview with *Fortune* magazine in 2003, Roth told the reporter: "Last year, we made over 5,000 compounds. Only half a dozen of them will make it to clinical trials. I'm spending a lot of time trying to understand the failures so that we can increase our odds of success."[34] With the popularity of Lipitor, honors deservingly followed. He was bestowed with the 1997 Warner-Lambert Chairman's Distinguished Scientific Achievement Award. At the award ceremony he gave detailed reminiscences of the discovery and development of Lipitor. In 1999, the lawyers of the New York Intellectual Property Law Association presented their Inventor of the Year Award to Roth for his invention of Lipitor. In 2003 Roth had two awards bestowed on him: the American Chemical Society Award for Creative Invention and the Gustavus John Esselen Award for Chemistry in the Public Service. More gratifying to Roth, when he moved about his small town of Plymouth, Michigan, it was not unusual for people to corner him in line at the supermarket or at a dinner party and offer their gratitude. When I talked to him in 2006, he mentioned that his happiest moment

Figure 4.3 Bruce D. Roth.

was when a fellow churchgoer stopped him in the parking lot, thanking him for saving his life.

Ironically, in 2001, Roth went to see his physician for a routine checkup and the blood work data came back, indicating that his LDL-cholesterol level was 160 mg/dL. While Akira Endo chose exercise and diet in place of Mevacor, Roth began taking his own invention, Lipitor, at the 10-mg dose. His LDL-cholesterol levels rapidly went down to 100 mg/dL, and at times it was as low as 74 mg/dL. His favorite story he told me was about Robert L. Smith, the inventor of Zocor. Smith also takes his own invention, Zocor, but his results are not as impressive!

In January 2007, when Pfizer announced closure of its Ann Arbor laboratories,[35] Bruce Roth accepted a position as a senior director of small-molecule chemistry at Genentech Inc. in South San Francisco.

CHAPTER 5

Development of Lipitor

Checkered Life

From Sankyo's experience, Roger Newton and his pharmacology colleagues at Parke-Davis learned that the rat is not a good animal model for cholesterol-lowering drugs because the rat's liver is able to rapidly compensate for the drop of cholesterol. So, for initial in vivo efficacy studies, Newton's team chose guinea pigs and divided them into two sets: one normal group and the other a chow-fed group. The pharmacologists then gave different groups of guinea pigs Merck's Mevacor (lovastatin) as a reference or CI-981 in one of three different doses. When the results were tabulated, Parke-Davis scientists were underwhelmed: not only was CI-981 no better than lovastatin in lowering cholesterol, but it seemed even a little inferior. However, they continued with the animal studies with a higher species—the dog. They used cholestyramine-primed dogs since the plasma cholesterol-lowering effects became more pronounced after the dogs had been fed with the cholesterol-lowering resin. Unfortunately, again in dogs, CI-981 showed LDL-cholesterol–lowering efficacy about the same as that of lovastatin at the doses tested. Upon further scrutiny of the data, the only advantage that the pharmacologists could find was that CI-981 lowered triglyceride levels slightly better (20–40%) than lovastatin in those animal models.[1] In 1988, Mevacor had already been on the market for an entire year. Merck's Zocor and Bristol-Myers Squibb's Pravachol were poised to be launched in 1991, and Sandoz's Lescol would be ready to launch in 1994. In addition, Rhône-Poulenc Rorer's dalvastatin was at a similar stage as CI-981 in development. Either dalvastatin or CI-981 would have become the fifth of its kind on the

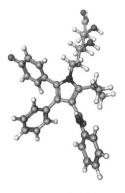

Figure 5.1 Molecular structure of Lipitor.

market. Parke-Davis believed that they would need differentiation in order to successfully sell its drug. After all, CI-981 had to be different from, hopefully better than, existing drugs, for them to make a good profit.

Parke-Davis's belief that they needed to differentiate CI-981 from existing statins had its root in another Parke-Davis drug, Accupril (quinapril), an ACE inhibitor poised to launch in 1991. Because Accupril was a "me-too" drug and did not have significant differentiation from other ACE (angiotensin-converting enzyme) inhibitors, it was not expected to do well on the market. The last thing that Parke-Davis management wanted was a repeat of the Accupril story.

ACE inhibitors are widely used to treat hypertension, congestive heart failure, and heart attacks. The first ACE inhibitor, captopril, was discovered by biochemist David Cushman and organic chemist Miguel A. Ondetti in the 1970s at the Squibb Institute,[2] which has marketed captopril under the brand name Capoten since 1978.

The discovery of captopril was a superb achievement in the history of drug research. Not only was it a novel mechanism for drug discovery, but it also inspired future applications of so-called rational drug design by taking advantage of the structures of the target molecule. The novel mechanism of ACE inhibitors and the success of captopril triggered a flurry of research in the pharmaceutical industry.[3] Captopril possessed a trio of side effects attributed to a thiol functionality: bone marrow growth suppression (due to a decrease in circulating white blood cells), skin rash, and diminution or loss of taste perception. Trying to make a better ACE inhibitor by improving upon captopril, a group of Merck scientists led by Arthur A. Patchett replaced the thiol group with a carboxylate. They

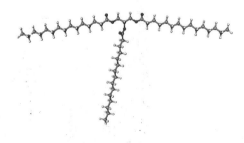

Figure 5.2 Triglyceride, like cholesterol, is also a lipid.

eventually succeeded with enalapril. Although it was a "me-too" drug, it was better absorbed by the stomach than captopril and it was also devoid of the side effects associated with the thiol group. In 1985, Merck successfully completed clinical trials, gained FDA approval, and sold enalapril under the trade name Vasotec, which became their first billion-dollar drug in 1988. A year earlier, Merck's lisinopril (Zestril) also reached the market. Lisinopril did not offer significant advantages over enalapril in efficacy, except once-a-day dosing. It helped increase patient compliance and was thus advantageous—it is surprising how low patient compliance is for drugs that have to be taken more than once a day. With their two ACE inhibitors, Merck dominated the field even though other pharmaceutical companies introduced their own versions of ACE inhibitors.[3] Roche's cilazapril (Inhibace) entered in 1989. In 1991 four new ACE inhibitors were approved by the FDA: Hoechst's ramipril (Altace), Squibb's fosinopril (Fozitec), Ciba-Geigy's benazepril (Lotensin), and Parke-Davis's quinapril (Accupril). They all have the same mechanism of action and similar side-effect profiles but differ in such areas as their relative potency, pharmacokinetics (what the body does to the drug), duration of action, and tissue distribution.

Sylvester Klutchkow, a chemist at Parke-Davis, invented quinapril in 1982.[4] The structure of quinapril bears a striking similarity to Merck's enalapril: three-quarters of the molecule is the same. The only difference is that quinapril has a tetrahydroquinoline in place of enalapril's pyrrolidine. However, Parke-Davis was able to patent it because Parke-Davis's quinapril had a six-membered ring and Merck's enalapril had a five-membered ring. After FDA approval in November 1991, Parke-Davis marketed quinapril under the trade name Accupril. Unfortunately, because Accupril was the sixth of its kind to enter the market and lacked significant differentiation from other ACE inhibitors, it held only a very small market share of 2.14%. Accupril's annual sales peaked at $706 million in 2003, and it has been available as a cheap generic drug since 2004.

In late 1989 and early 1990, management's indecision on CI-981, their most promising candidate, was causing a lot of anxiety to the Parke-Davis statin team. Without a human clinical trial, they would never know what they had in their hands. Indeed, animal studies comparing it with Mevacor were not impressive. But humans often respond differently than animals to certain drugs. In the first quarter of 1990, Newton made an effort to convince the management to apply for Investigational New Drug (IND) status for CI-981 so they could advance it to phase I clinical trials. To do so, they had to convince a committee chaired by Ronnie M. Cresswell, the chairman and president of Parke-Davis research and development. Cresswell, a Scottish process chemist, was hired by Parke-Davis in the late 1980s. In addition to the bigwigs in research and development, several marketing people also descended from the Warner-Lambert headquarters in Morris Plains, New Jersey, to take part in the meeting. By then, the team had already spent eight years in the statin program, so a lot of investment was riding on this decision. With one statin (Mevacor) already on the market and four others (Merck's Zocor, Bristol-Myer Squibb's Pravachol, Sandoz's Lescol, and Rhône-Poulenc Rorer's dalvastatin) in late-stage human studies, some Parke-Davis managers feared that the potential payoff was too slim to justify further development of CI-981.

Newton and Bruce Roth attended the meeting, and Newton gave the presentation on behalf of the team. Both men knew what was at stake. When Newton ascended to the podium, he seemed unusually calm. He began by presenting the efficacy and safety data of CI-981 in animal models, admitting that there was not much differentiation from other statins. Then, he argued that humans *often* respond differently than animals and that the team really believed in the drug; otherwise, they would not ask the company to make further investments. After all, without clinical trials in humans, one would never find out how the drug would behave in humans.

At the end of his presentation, Newton made an impassioned plea for allowing CI-981 a shot at human trials. He argued: "We'd spent a lot of years of blood, sweat, and tears on this compound. Why would you spend all this money just to get to the edge of where you find out if it's viable?"[5] Suddenly, out of the blue, he started to sing his own version of Al Jolson's number "You Made Me Love You," falling to one knee while he crooned: "You've got to let us do the human tests. I know it's the right thing to do, and I'm begging you to do it."[6] The managers on the committee were initially somewhat startled by Newton's unconventional stunt, yet they all enjoyed a laugh with him. Seeing that his joke had broken the ice, Newton became serious and outlined a detailed calculation with the managers. Indeed, phase I trials might cost the company $1.2–1.5 million. But the cholesterol market was estimated to be $5–6 billion then (in 2006, it was more than $32 billion). Even if all four of the other statins in development made it to the market before CI-981 (one of them, Rhône-Poulenc Rorer's dalvastatin, ultimately did not) and CI-981 caught a share of 5% as the fifth player, the potential sales could still amount to about $300 million. Parke-Davis's best-seller in 1990, Lopid, had annual sales of $600 million at its peak. Sales of nearly half those of Lopid were not bad for a company of Parke-Davis's size. Therefore, CI-981 would still be profitable enough to warrant moving it forward.

Eventually, management was convinced by Newton's passion, conviction, and, most crucially, his data and arguments. The committee unanimously agreed to test CI-981 in humans. Today, few key players and executives would readily admit this, but the reality was that one of the largest motivating factors for the decision of moving CI-981 to human trials was that Parke-Davis's drug pipeline was relatively dry. It is chilling now to even contemplate that Lipitor might not have even reached the human trial stage if Parke-Davis had enough other drugs to develop at the time.

Chiral or Not, You Have to Make It First

Even with the green light from the committee, large quantities of CI-981 were still needed in order to embark on the phase I clinical trials. A drug's journey from the laboratory to the market is often long and arduous. Lipitor's journey was especially tumultuous—overcoming the synthetic challenge of the complex chiral molecule was just another step of its thousand-mile journey.

In the spring of 1988, the statin team made the decision to synthesize the enantiomerically pure form of the drug. The decision had scientific reasons in addition to logistical mandates. First of all, a chiral drug would obviate the unnecessary burden to the patient of having to metabolize the 50% less active material (the wrong enantiomer). In addition, it would give Parke-Davis an edge with superior activity to overcome the obvious disadvantage in a compound entering the market place potentially 10 years after the fungal metabolite-derived inhibitor, Mevacor. History proved that the decision to make the chiral drug was a wise one, for it indeed gave Parke-Davis an edge in the competitive field of statins. For instance, the lowest dose approved by the FDA for Lipitor was 10 mg/day. Without an optically pure drug, it would have been 20 mg/day. Scrutiny of other HMG-CoA reductase inhibitors in development in that period of time reveals that racemates did not do very well. Sandoz's fluvastatin (Lescol) did not do as well on the market as Lipitor, with annual sales of $734 million compared with Lipitor's $9.2 billion in 2003. Rhône-Poulenc Rorer's dalvastatin failed to even reach the market.

Due to the necessary division of labor, most chemists in drug companies generally specialize in one of two areas: drug research discovery (medicinal chemists) or drug development (process chemists). Normally, when a drug is discovered and moved into clinical development, the torch is passed on from medicinal chemists in drug discovery to process chemists in drug development. Because of the difference in their job descriptions, what they do is quite different, as well. Medicinal chemists' job is to discover drugs with desired pharmaceutical properties such as potency, efficacy, bioavailability, and safety. They only need to make enough to collect those data—usually only a few grams would suffice nowadays. As a consequence, they generally do not worry themselves with making a large quantity of the drug. On the other hand, process chemists' job is transforming the synthesis developed by medicinal chemists into a route that can be used in manufacturing facilities for large-scale production in kilogram quantities of the drug. More often than not, they have to invent new routes for the synthesis because the original discovery route is inadequate for large-scale production. In essence, they bridge the gap between the laboratory and the manufacturing plant, where tons of drugs are produced. As a consequence, they concern themselves greatly with the safety, scalability, reproducibility, environmental impact, cost of goods, and yields.

For Parke-Davis in those days, discovery chemistry was located in Ann Arbor, Michigan, and chemical development was sited in Holland, Michigan, about 150 miles west of Ann Arbor. Near the eastern shore of Lake Michigan, Holland is an immigrant town of Dutch descent, thus the town's name. When CI-981 was declared a lead compound, the chemistry was transferred from Ann Arbor to Holland. Donald E. Butler, the chemist in charge of the project, considered its synthesis one of the most challenging in process chemistry. Butler had been a medicinal chemist himself for 30 years and became a group leader of Parke-Davis's chemical development in Holland in the 1990s. His appraisal of the synthesis was right on. In fact, the Ann Arbor discovery team had difficulty even making enough CI-981 for the initial toxicology batch required for safety studies. As a general rule, one prefers to have a synthesis of fewer than 10 steps. One of the few drugs that required more than 10 steps to make was prostaglandin, made by Upjohn in Kalamazoo, Michigan—but because prostaglandins are extremely potent, very little of the drug is needed in each dose, so a long and tedious synthesis was acceptable.

The original synthesis of CI-981 developed by Roth was 17 steps long, clearly not amenable to large-scale process and manufacture. Due to the intricacy of synthesizing CI-981, Parke-Davis actually formed two teams to develop its process chemistry instead of the usual handoff from discovery to process. One team was led by Bruce Roth and consisted of chemists involved with the HMG-CoA reductase inhibitor program, including Alex Chucholowski, Bob Sliskovic, and W. Howard Roark. The development team in Holland, led by group leader Don Butler, consisted of Kelvin L. Baumann, Philip L. Brower, Carl F. Deering, Rex Jennings, Tung Le, Kenneth Mennen, Alan Millar, Thomas N. Nanninga (another group leader), Charles Palmer, and Robert A. Wade.

In time, the first chiral synthesis was developed in Ann Arbor by Alex Chucholowski, who worked under Roth's direction. He employed the chiral acetate enolate chemistry developed by Manfred Braun of University of Düsseldorf in Germany. Chucholowski's asymmetric synthesis worked very well on a small scale in the laboratory, producing CI-981 with greater than 99% enantiomeric purity. The team was ecstatic. Unfortunately, when Chucholowski's chemistry was applied to large-scale production in Holland, it was a complete failure because it was not robust enough to make the quantity that was required, and it involved too many low-temperature reactions. So the chemical development team in Holland went back to the drawing board to develop an alternative chiral synthesis.

Ironically, the first challenge that confronted the team was not the chiral synthesis, but scaling up the achiral part of the molecule. Alan Millar did something that no medicinal chemist had the luxury to do: he spent *an entire year* looking at one transformation, the Paal-Knorr synthesis of pyrroles (two German chemists, C. Paal and L. Knorr, discovered the reaction in 1885). Although the reaction had worked well enough for the medicinal chemists, the process chemists were initially unable to achieve the level of efficiency that was demanded of production-scale chemical synthesis. Millar varied numerous reaction parameters, such as reaction stoichiometry, acid catalyst, temperature, and solvent. In the end, he was able to make the fully substituted pyrrole core in one step with a 75% yield using the Paal-Knorr reaction of a highly substituted ketone.

In the late 1980s and early 1990s, the field of asymmetric synthesis was not as mature as it is today. "The chemistry needed to get the right relationship of the atoms was just becoming available," commented James Zeller, team leader for the Lipitor drug development active pharmaceutical ingredient team, in a later interview.[7] Meanwhile, Tom Nanninga tried to develop a chiral synthesis of the side chain. He spent months experimenting with various processes, but the selectivity never reached a level acceptable for production. Ultimately, the chemical development department purchased a reactor that could be cooled to −80°C by using liquid nitrogen, which allowed the high selectivity needed.[8]

Tung Le, a Vietnamese American who was working under Nanninga, took on the task of optimizing the chiral acetate enolate chemistry developed by Chucholowski. He made an important contribution to the project by discovering that the reaction could be carried out without using protecting groups for a hydroxyl group. Typically, using protecting groups in organic synthesis is a fact of life. However, for process chemistry, protecting groups necessitate at least two additional steps—one to add the protecting group, and another to remove it—which translates to a tremendous amount of added cost and labor. Le's route without a protecting group was therefore a great boon to the process.

Eventually, the development team optimized conditions for each step. By 1990, enough CI-981 was synthesized to begin the human trials.

Clinical Trials—Divine Providence?

With enough CI-981 in hand, and with the blessing of management, Parke-Davis was at last ready to move CI-981 into phase I clinical trials, also known as "first in humans" trials. In order to test a new drug on humans, one cannot simply gather a bunch of people and try the new drug in them—although that was exactly what pioneer scientists used to do in the early twentieth century.

In 1910, when Paul Ehrlich discovered arsphenamine to treat syphilis, he and his Japanese associate Sachachio Hata first tested the drug in dogs. After finding that arsphenamine killed the syphilis microbes without killing the dogs, Ehrlich wanted to try it on himself. Since he was already 56 years old and in poor health, two of his assistants volunteered instead to be human guinea pigs, and arsphenamine was subsequently found to be relatively safe. Arsphenamine had a tremendous impact in fighting syphilis, wiping out half of all syphilis infections in Europe in just five years.

In the early 1930s, Gerhard Domagk, head of the bacteriology laboratory of I. G. Farben in Germany, experimented with different dyes in mice, in search of antibacterial drugs. He discovered that a dye $2',4'$-diaminoazobenzene-4-sulfonamide, later branded as Prontosil, was effective at killing bacteria in mice with limited toxic side effects. In November 1932, Domagk's six-year-old daughter, Hildegarde, became ill with an infection from the prick of an embroidery needle contaminated with streptococcal bacteria. The infection quickly spread to her lymph nodes, and blood poisoning became severe. Other doctors recommended amputation of her arm, but even that drastic measure would not offer a good chance of survival. Near death and unresponsive to other treatments, Hildegarde was injected by Domagk himself with a large dose of Prontosil. She made a miraculous recovery, and thus the first sulfa drug was born.

Unfortunately, not all human trials had happy endings. In late 1935, the real active ingredient in Prontosil was discovered to be sulfanilamide. Because of sulfanilamide's insolubility in ethanol, there was no syrup or liquid form, which would have been easier for children to take. The S. E. Massengill Company in Bristol, Tennessee, dissolved sulfanilamide in diethylene glycol and water, added some pink raspberry flavor, and branded the concoction "Elixir Sulfanilamide." With no check for toxicity and no clinical trials, Massengill's 200 eager salesmen rushed to sell it to doctors and pharmacies all across the country. Although sulfanilamide was thoroughly tested for safety in animals and humans and was proven to be reasonably safe, diethylene glycol is very toxic. In the end, 107 patients died of kidney failure.

Elixir Sulfanilamide was one of the worst medical blunders in American history, but the most notorious unethical clinical trial was probably the Tuskegee Syphilis Study, officially known as the "Tuskegee Study of Untreated Syphilis in Negro Males." In 1932, the U.S. Public Health Services (PHS), in collaboration with the Tuskegee Institute (now Tuskegee University), recruited 399 poor and illiterate African Americans with syphilis, and 201 African American control subjects not infected with syphilis, in Macon County, Alabama.[9] Not only were the syphilis patients not informed of their infection (instead, they were told they were being treated for "bad blood," a local term for a range of diseases including anemia and fatigue), but they were also not given any treatment, even after penicillin became a standard treatment for syphilis in 1947. The study went on for 40 years until in 1972 conscientious doctors, especially Dr. Peter Buxtun, raised ethical concerns. This racist experiment was terminated in October of that year when it was found to be "ethically unjustified" by an advisory panel. By then, more than 100 of the subjects had died of tertiary syphilis, 40 of their wives were infected, and 19 of their children were born with congenital syphilis. In 1997, President Bill Clinton officially apologized to the victims on behalf of the nation: "What was done cannot be undone. But we can end the silence. We can stop turning our heads away. We can look at you in the eyes and finally say, on behalf of the American people: what the United States government did was shameful. And I am sorry."[10] Indeed, it is appalling that such flagrant immorality took place under the auspices of the U.S. government.

Nowadays, such unethical practices are strictly forbidden in the United States and most developed countries. Bioethics is strongly emphasized with regard to medical research, including clinical trials. Drug testing is now highly regulated by many governments. For example, the FDA requires the sponsor of a drug trial to submit an Investigational New Drug (IND) application, for which the safety of the drug must be well established in animal models.

In 1990, Parke-Davis had compiled all of the relevant pharmacology, pharmacokinetics, drug metabolism, and safety data for CI-981 and submitted it to the FDA. At that time, Mevacor had already been on the market for more than a year. Furthermore, the INDs for Zocor, Pravachol, and Lescol had been approved by the FDA several years prior to Parke-Davis's IND filing for CI-981. Therefore, the FDA was very familiar with statins as HMG-CoA inhibitors for the treatment of hypercholesterolemia. Not surprisingly, the IND for CI-981 was promptly approved.

While Roger Newton was still somewhat involved with phase I trials, the discovery team responsible for the birth of CI-981 passed the torch to Parke-Davis's development team. The phase I studies were carried out in Parke-Davis's clinical research unit (CRU), located on Parke-Davis's sprawling campus in Ann Arbor, Michigan. This CRU was managed by Director Ralph Stein and Vice President Alan Sedman in Parke-Davis's clinical development department. The key purposes of most phase I trials are to gauge the safety profile of the drug, how the body reacts to the drug (pharmacokinetics, or PK), and how the drug acts in the body (pharmacodynamics, or PD). The CI-981 development team recruited 24 healthy volunteers internally from Parke-Davis's own ranks. Those volunteers were all men because it would be unethical to test a drug with an unknown safety profile in women who could become pregnant during the trials. Not only were the 24 healthy male volunteers compensated well for their participation, but they were also given a thorough and stringent physical examination, including blood work, ECG, and urinalysis. The men were sequestered in the CRU so that the clinicians could carefully monitor their reaction to the drug. This also made it easier for them to provide emergency care if anything went awry. Before the 1970s, drug firms routinely recruited male prisoners for phase I trials since those prisoners were confined and it was a lot easier to track their clinical changes. As a matter of fact, until 1970, Parke-Davis's phase I trial site was located in the state prison in Jackson, Michigan, a town 20 miles west of Ann Arbor. This kind of practice does not occur today, because prisoners are considered unable to provide informed consent that is freely given and independent of any coercion.

As in any typical phase I trial, each of the 24 healthy volunteers was initially given one of the low doses (2.5, 5, or 10 mg) for one day. This trial was called the single-dose/single-day trial, and it was designated as the 981-1 trial. The purpose was to make sure that the drug in question did not have any acute toxicity that could cause adverse events from a single dose. Although it is extremely rare for that kind of event to take place, it does occur occasionally. For instance, on March 13, 2006, four volunteers in London who tested TGN1412, an experimental monoclonal antibody, suffered terrible consequences even though tests had shown it was safe in animals.[11] TGN1412 was developed by a 15-person biotechnology company TeGenero Immuno Therapeutics in Germany for the treatment of leukemia and rheumatoid arthritis and worked by lowering the levels of CD-28. Six healthy male volunteers, either poor immigrant students or unemployed, were promised £2,000 to take part in the trial. Although given only a subclinical dose of 0.1 mg/kg, some 500 times lower than the dose found safe in animals, shockingly, the four who received the drug experienced catastrophic immune responses. All were hospitalized, and they suffered multiple organ dysfunctions, some requiring organ transplants to save their lives. Meanwhile, the two healthy male volunteers given placebo were perfectly fine.

Thankfully, no acute toxicity was observed for CI-981's single-dose/single-day 981-1 trial. With the safety profile secured for lower doses, Stein and Sedman began trial 981-2, which was also a single-dose/single-day trial but tested larger doses of 20, 40, 80, and 120 mg. The purpose of trial 981-2 was to gauge how high they could increase the dose and still observe an acceptable safety profile. Although volunteers given doses of 2.5, 5, 10, 20, 40, and 80 mg experienced no adverse effects, intolerance was observed for men on the 120-mg dose—some liver toxicity manifested with elevated liver enzyme levels. The team backed off the 120-mg dose and determined the upper limit of the dose would be 80 mg/day.

The development team then moved on to the second component of phase I trials, where the 24 healthy volunteers were given CI-981 dosages of 2.5 to 80 mg/day for 2 weeks,[12] designated as trial 981-3. One thing was on their side—when carrying out a clinical trial for hypercholesterolemia, you need to measure only the levels of LDL and HDL cholesterol, triglycerides, and other lipids, whose changes would only take a couple of weeks to manifest. If the clinical trials were for Alzheimer's disease or osteoarthritis, the effects for even the most effective drugs would take months if not years to manifest. Therefore, the results were obtained quickly for CI-981 at Parke-Davis's CRU. When the final results were tabulated, everybody was impressed. The drug performed far better for humans than for animals! Not only were there no significant safety issues among all volunteers, but their cholesterol levels were significantly lowered. Volunteers on the 10 mg/day dosage had their average LDL-cholesterol levels drop 38%, which was as good as, if not better than, most known statins. As a comparison, Mevacor, at its highest FDA-recommended dose of 40 mg, lowered the average LDL-cholesterol levels by approximately 39%.

One episode gave the CI-981 development team a bit of a scare. While tabulating the analytical data, the team noticed that HDL-cholesterol levels had also dropped. That would be bad news—since HDL-cholesterol is the "good cholesterol," which has protective effects on the heart, lowering its levels would be a drawback for the drug. Fortunately, upon closer scrutiny, the team discovered that the data had been generated using a nonstandard assay carried out in Parke-Davis's toxicology laboratories. After switching to the well-calibrated precipitation assay, CI-981 was instead found to boost HDL-cholesterol levels, consistent with other members of the statin class.

More impressive, volunteers given 80-mg doses had their LDL-cholesterol levels drop 58%, a magnitude of decline never seen in any other statins at the time! The efficacy of CI-981 blew all competitors out of water. Ironically, when the team reported this to management, they woefully commented that everything seemed positive with only one downside: LDL-cholesterol levels were lowered so rapidly and so precipitously that they were afraid of unknown negative ramifications. Later trials were soon proved their worries to be unfounded, echoing cardiologists' mantra: lower is better.

Again, humans have different responses to drugs than do animals. However, bioethics strictly forbids testing drugs using humans as guinea pigs without first gathering extensive safety data from animals. Therefore, scientists in government, academia, and industry have spent tremendous resources in developing animal models for testing the efficacy and safety of drugs. This actually creates a dilemma for the scientists: although the drugs are developed for humans, they have no choice but to use animal models to gauge how those drugs are likely to behave in humans. Because of the discrepancy between humans and animals in the response to drugs, from time to time one gets a surprise. The dichotomy also makes drug discovery very expensive. Sometimes, regardless of how well a drug performs in animal models, it does not work for humans. Nowadays, clinical trials in humans could cost as much as $1 billion, so the discrepant responses between humans and animals could easily bankrupt a small or even a medium-sized pharmaceutical company if the wrong decision was made.

For CI-981, which would become Lipitor in time, the discrepancy between humans and animals meant future revenues of tens of billions of dollars per year. When the phase I study data were reported, gone was the worry of some, but not all, in the company about the little word "differentiation"—they had all the differentiation they needed. CI-981 was not only different, it could be much better and safer.

Canter and the FDA

The impressive results seen in the phase I volunteers dispelled the doubts that many had at the time. Naysayers in the company became silent, and everyone involved began to feel the excitement: maybe, just maybe, they had a winner on their hands! As soon as phase I studies ended, Parke-Davis embarked on phase II investigations. In drug development, while phase I trials are largely used to gauge the toxicity and how the drug and human body interact with each other, phase II clinical trials are intended to measure the efficacy of the drug—in other words, whether it works for the proposed illness. Whereas phase I trials normally need one or two dozen healthy volunteers, phase II trials generally involve a larger number of patients. While phase I clinical trials are known as "first in humans" trials, phase II clinical trials are known as "first in patients" trials. At the onset of phase II studies of CI-981, another key player in Lipitor's fate emerged: David Canter, senior director and head of the cardiovascular clinical group at Parke-Davis's drug development department. While the majority of Canter's group were occupied with the clinical trials of Accupril, he hired Donald M. Black, from a company in Cincinnati, Ohio, as director of cardiovascular disease specifically to run the phase II and III clinical trials for CI-981. The phase II clinical trials for CI-981 began with Don Black as the project leader. His team included Rebecca G. Bakker-Arkema, Harry E. Haber, and James W. Nawrocki. If one were to view Bruce Roth and Roger Newton as Lipitor's parents, giving birth to Lipitor, then David Canter, Don Black, and their colleagues would be the nannies, shepherding CI-981 through the clinical trials and passing the baton to the FDA.

Canter, born on November 15, 1953, in Reading, England, spent much of his youth in boarding schools. Since his father was a lieutenant commander of the Royal Navy, he had opportunities to live in Hong Kong and Singapore. These overseas experiences opened his eyes to the wide world outside the tiny British Isles. In 1971, Canter went to the University of Cambridge, one of the oldest universities in the world and one of the largest in the United Kingdom. Some of Cambridge's famous alumni include Isaac Newton, William Harvey, Charles Darwin, Francis Crick, and James Watson. After obtaining his B.A. in natural sciences in 1974, Canter enrolled at the University of Liverpool, where he earned a degree of M.B. Ch.B. (Medicinæ Baccalaureus et Baccalaureus Chirurgiæ, equivalent to America's M.D.) in 1979. Jobs during the late 1970s in England were not plentiful. After six months of internship at the University of Liverpool, Canter took on a surgical residency at the National Health Service. The hours were brutal (120 hours per week!) and the pay was meager (residents were paid in full for only 40 hours per week and a third for the remaining hours). When his residency ended, he promptly lost his employment as well. When Canter showed up at the unemployment office, he was asked if he had worked during the last week. After he answered; "Yes, I was a surgeon at the Hospital performing surgery a week ago," and everyone in the unemployment office promptly became dead silent.[12]

In any case, unemployment gave Canter time to ponder what he wanted for his life, and he determined to change his life dramatically. He applied for and was offered a position with Parke-Davis in 1984. He started as an associate director in clinical development in Pontypool, South Wales, and moved to Parke-Davis's headquarters in Ann Arbor, Michigan, in 1986 as a director in the cardiovascular diseases department. He managed the congestive heart failure clinical trials of Accupril and quickly ascended to senior director in 1989. When the phase II trials for CI-981 began, Canter became a champion for the drug, steering it through the twists and turns until the New Drug Application (NDA) submission to the FDA, and became a vice president of drug development in 1992.

Phase II trials for CI-981 involved about 81 patients. Don Black's team enlisted clinicians from outside Parke-Davis, including Stuart R. Weiss in San Diego, California; Michael H. Davidson in Chicago, Illinois; Dennis L. Sprecher in Cincinnati, Ohio; Sherwyn L. Schwartz in San Antonio, Texas; Paul J. Lupien in Quebec, Canada; and Peter H. Jones in Houston, Texas. Eighty-one patients were first observed for an 8-week placebo-baseline dietary phase, and they were then randomly assigned to receive either placebo or 2.5, 5, 10, 20, 40, or 80 mg of atorvastatin once daily for six weeks. The clinicians observed that the plasma LDL-cholesterol reduction in patients ranged from 25% to 61% with the minimum dose of 2.5 mg to the maximum dose of 80 mg CI-981 taken once daily. Reductions of this magnitude had never before been observed for a single drug. In other words, CI-981 outperformed every single statin on the market at the time. Equivalent efficacy was achieved only when combination drug therapy was used. In this study, CI-981 was well tolerated by hyperlipidemic patients, had an acceptable safety profile, and provided greater reduction in cholesterol levels than any other previously reported statin.[13]

At the end of the first phase II trial—the 981-4 trial (-4 signifies that it was the fourth human trial of the drug)—the results were tabulated as changes in LDL

Figure 5.3 David Canter, 2001.

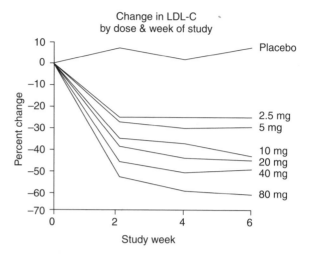

Figure 5.4 The 981-4 Trial: Changes in LDL-cholesterol by dose and week of study © Pfizer.

cholesterol by dose and week of study. This chart for the 981-4 trials convinced everyone that CI-981 really worked. To drive the message home, Canter and Black gathered literature data and put together a chart that compared CI-981 with all known statins at the time. Now famously known as "the curves," the chart clearly showed the superiority of CI-981—in terms of efficacy, it was far superior to all existing statins! Parke-Davis's sales department was so impressed with "the curves" that they asked the company to actually carry out a clinical trial to compare atorvastain with competitors during phase III studies, just to reproduce it. As a consequence, they could better differentiate CI-981 from other statins on the market in their sales pitch to the physicians. Their investment was handsomely repaid in the future. In the end, the FDA even allowed Parke-Davis to use "the curves" on Lipitor's package insert (the pamphlet that accompanies each filled prescription).

In 1994, armed with these impressive data, about 10 Parke-Davis representatives flew to Rockville, Maryland, and met with FDA staffers to request the FDA grant CI-981 priority status.[14] Leading the Parke-Davis team was Irwin Martin, vice president of regulatory affairs, along with Canter and Black. The meeting lasted more than an hour. The FDA staffers were not impressed—after all, four statins (Mevacor, Zocor, Pravachol, and Lescol) were already on the market, and there was no urgency for a fifth one. However, they mentioned that the agency could grant CI-981 priority status if it was proven efficacious for a disease that had an unmet

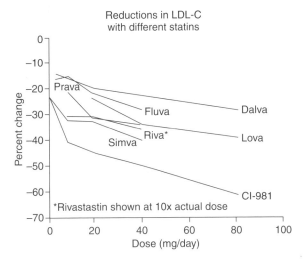

Figure 5.5 "The Curves": Reduction in LDL-cholesterol with different statins © Pfizer.

medical need, such as familial hypercholesterolemia (FH), a genetic disease where the patients' cholesterol levels are extraordinarily high. The homozygous form of FH is severe but rare, striking only one in a million—more prevalent in Quebec, Lebanon, and South Africa, the disease affects no more than 100 patients in the United States. These patients have cholesterol levels greater than 1,000 mg/dL at birth, often begin to have heart attacks in their childhood, and have an average life expectancy of 20 years. Although many drugs, including the earlier statins, had been tried as treatment for these homozygous FH children, none had any significant efficacy. Parke-Davis creatively designed a clinical trial involving the homozygous FH children, and atorvastatin was proven to be efficacious. The FDA then granted CI-981 priority status, which gave Lipitor an early coming of age and possibly half a billion dollars in sales.

The phase II studies involved six separate trials. At the meeting held at the end of phase II, the FDA representatives commented that the efficacy of CI-981 was not an issue at all, but the safety could be. As a consequence, the development team had to prove to the agency that CI-981 was not only efficacious but also safe, which could only be achieved through phase III clinical trials that involved thousands of patients. In most phase III trials, the experimental drug or treatment is given to large groups of people (1,000–10,000) to confirm its effectiveness, monitor side effects, compare it to commonly used treatments, and collect information that will allow the experimental drug or treatment to be used safely.

A drug advancing to phase III earns the distinction of being endowed with a generic name. The generic name for CI-981 was relatively painless to come by. The U.S. Adopted Names (USAN) Council, sponsored by the American Medical Association, is in charge of sanctioning generic names, also known as USAN names. To organize drugs with the same function, the USAN Council normally defines a "suffix" for each class of drugs. For instance, "-vastatin" was chosen for drugs that lower cholesterol and inhibit the buildup of plaque in arteries. Since CI-981 belongs to the statin class, all that was left to do was add a certain prefix to the stem "-vastatin." Parke-Davis initially contemplated of giving CI-981 the generic name "torvastatin." But somebody in the company felt that the drug would have more exposure if it was listed at the top of tables in drug formularies, where drugs in the same class are invariably listed alphabetically by generic name. Parke-Davis settled on the generic name "atorvastatin," and this was duly approved by the USAN Council. Atorvastatin calcium does always appear prominently at the top of tables of statins, rather than buried somewhere deeper down. Of course, a good drug is the key to success—a terrible drug would never become popular even if its generic name began with "o"!

David Canter delineated an "atorvastatin strategic plan." In that plan, he bravely proposed to test atorvastatin head-to-head with other statins on the market, including Mevacor, Zocor, and Pravachol. That was a very tricky proposition. First of all, the FDA did not require Parke-Davis to prove that atorvastatin was superior to the existing statins in order to approve the drug. Second, if atorvastatin tested as inferior to other statins, it would be difficult for Parke-Davis to sell it. As we will see in chapter 6, the PROVE-IT trials got Bristol-Myers Squibb into just such a situation for Pravachol in 2005.

Canter's atorvastatin strategic plan also defined the target population and appropriate dose range. Since the most obvious distinguishing characteristic of atorvastatin was the reduction of LDL cholesterol by 40–60% at doses from 10 to 80 mg/day, Canter proposed that the starting dose be 10 mg/day. Since the 10-mg dose already had both strong LDL-cholesterol-lowering properties and increased potency compared to other approved statins, particularly Zocor, the minimal dose with fewest adverse side effects would be well accepted by physicians. Furthermore, intervention trials had shown that a 30–35% drop of LDL cholesterol led to a significant drop in coronary heart disease, and 10 mg of atorvastatin was already capable of lowering cholesterol levels by 38–40%. Although 2.5- and 5-mg doses were tested, these doses did not provide an impressive enough drop in cholesterol levels. Most physicians are comfortable prescribing doses twice or even four times that of the starting dose, which would be 5 and 10 mg, respectively, if 2.5 mg were chosen as the starting dose. As a consequence, although the 2.5-mg dose was good enough to gain FDA approval, if that were listed as the starting dose, a physician would not be comfortable prescribing doses of 20 and 40 mg. With 10 mg as the starting dose, they would not hesitate with 20 and 40 mg doses because the highest approved dose (80 mg) was already demonstrated to be safe. Finally, since many patients did not require a spectacular reduction of LDL cholesterol, Canter proposed that Parke-Davis use the triglyceride-lowering properties (20–40%) of atorvastatin as an additional point of differentiation. This proved to be prescient. When asked, "Why should I use atorvastatin?" the sales representative could reply because of its superb efficacy at lowering both cholesterol and triglyceride levels. In hindsight, setting the starting dose at 10 mg was a brilliant decision.

The phase III clinical trials of atorvastatin recruited 3,802 patients who were divided into two groups. The first group of 2,502 patients was given atorvastatin in different doses. The second group, consisting of 1,300 patients, was further divided into several control groups and given Mevacor, Zocor, Pravachol (742 patients were given those three statins), Lopid, and cholestyramine. Normally, most phase III studies include a placebo group, but at the time the benefits of lowering cholesterol had become so well established that Parke-Davis decided to forgo the placebo group. Although the lack of a placebo group created some complications in tabulating the efficacy and safety data of atorvastatin, the company felt ethically more comfortable because all participants in the trials were given a cholesterol-lowering drug in one form or another.

By the end of 1993, data from 21 clinical trials of atorvastatin conducted in the United States and international community- and university-based research centers began to pour into Ann Arbor. In those trials, patients with lipid disorders were given atorvastatin at dosage of 10 to 80 mg/day for 4 weeks to more than 24 months. The development team, including Rebecca Bakker-Arkema,[15] Jim Nawrocki,[16] and Don Black,[17] spent all their time processing and analyzing the data. In the end, each of the three wrote a review on atorvastatin's safety profile. After pooling data from 2,502 patients from the 21 clinical trials, as well as 23 then ongoing clinical trials involving 1,769 patients, they unanimously concluded that atorvastatin's safety profile was consistent with all of the statins tested (Mevacor, Zocor, and Pravachol) and was similar to that seen in other compounds of this class.

What's in a Name?

While phase III trials for atorvastatin were drawing to an end, Parke-Davis set out to identify a trade name for atorvastatin calcium in anticipation of its marketing. Discovering and developing a drug is a daunting task, as we have witnessed through atorvastatin's ups and downs. Naming a drug, as it turns out, is not such a smooth ride, either. One might assume that a good drug could sell by merit alone by quoting Shakespeare: "What's in a name? That which we call a rose by any other name would smell as sweet." But Shakespeare would be wrong if the name was attached to a modern drug. For one thing, the drug maker would like a trade name that is easy to pronounce and thus easier for the pharmacist and general public to remember. Yet there are so many drugs currently on the market that one has to be extremely careful to choose a name that will not be easily confused with others, which could have dire consequences. Naming a drug is big business—firms exist solely for this purpose. Coming up with an appropriate name for a drug could cost a drug firm as much as a million dollars.

When the time came for Parke-Davis to select a trade name for atorvastatin, the company first decided to solicit proposals from its own employees. Thousands volunteered. Parke-Davis chose several possible brand names and forwarded them to the Brand Institute, a branding company based in Miami, Florida. The Brand Institute first checked the names against their databank to ensure that those names would not be infringing others' copyrighted trademarks. Second, they confirmed that the names would not mean something vulgar or misleading. For instance, General Motors's Chevrolet Nova cars did not sell well in Spain, Mexico, and other Spanish-speaking countries—it turned out that "no va" means literally "doesn't go" in Spanish!

With a group of possible trade names on hand, the Brand Institute hired a group of physicians and pharmacists for mock prescriptions and filling of the scripts in order to ensure that the names were not easily misread or misheard. After a year of intensive research, Lipitor emerged as the trade name for atorvastatin calcium. It is a brilliant name; not only is it easy to pronounce, but "Lipitor" contains part of the word "lipid" and part of the generic name "atorvastatin." The name was suggested by Mi Kyung Dong, a Korean American who at the time worked as a project manager in Parke-Davis's drug development group. When she saw the contest from the marketing planning group for the atorvastatin trade name suggestions, she decided to submit "Lipitor." The major concept she thought about was "take away lipid." To her surprise, marketing people selected "Lipitor" as the trade name for atorvastatin a couple of years later. In December 1998, representing management, Anthony Wild,

Figure 5.6 Mi Kyung Dong, 1997.

the senior president of Warner-Lambert in charge of research and development, handed Mi Dong a check for $20,000 in front of an audience of hundreds of Parke-Davis employees. He said, "I understand that Mi Dong saved Parke-Davis a lot of money by coming up with the name Lipitor; however, this check is probably a little bit less than what we would have paid a trademark research firm to come up with a name."[18] The company definitely made a steal because the bill could have easily exceeded $1 million if the name had been created by an outside firm. And Lipitor is now a household name.

Parke-Davis filed the NDA for Lipitor in June 1996.

A Marketing Partner

Despite having splendid results from clinical trials, and despite having a catchy trade name, Lipitor was still viewed with uncertainty by some at Parke-Davis about its market potential. Indeed, Lipitor launched in 1997 was the fifth statin on the market, behind Merck's Mevacor and Zocor, launched in 1987 and 1991, respectively. Bristol-Myers Squibb's Pravachol was launched in 1991, and Sandoz's Lescol in 1994. Because Lipitor was a "Johnny come lately," the initial projection of annual U.S. sales was a meager $300 million according to their 1995 calculation. With that kind of dismal forecast, marketing people at Warner-Lambert, Parke-Davis's parent company, were convinced that the only way to achieve success on the market was to pair up with a sales-savvy comarketing partner.

At the beginning of phase I studies, the development team visited Rhône-Poulenc Rorer, whose dalvastatin was also in early-stage human studies. Parke-Davis proposed a comarketing deal, and Rhône-Poulenc Rorer turned down the request because their own statin dalvastatin was more advanced in human clinical trials than was CI-981. Interestingly, after CI-981's impressive results in humans became public knowledge, Rhône-Poulenc Rorer, with their dalvastatin having flopped in clinical studies, approached Parke-Davis about the original offer. But it was no longer on the table.

The development team then solicited the interest of Wyeth-Ayerst Pharmaceuticals. David Canter vividly recalled their visit. After Parke-Davis showed all their cards, Wyeth's development department commented that the value of a statin fifth on the market was seriously limited. Needless to say, Parke-Davis came back to Ann Arbor empty-handed. Afterward, they assembled three lists of possible comarketing partners. The A-list included Fujisawa in Japan, Imperial Chemical Industries Ltd. in the United Kingdom, and Marion-Merrill Dow, Eli Lilly, Pfizer, and Syntex, all in the United States. The B-list consisted of the British firm Glaxo-Welcome, German firm E. Merck, Swiss firms Ciba-Geigy and Roche, French company Rhône-Poulenc Rorer, and the U.S. company Searle. Finally, the C-list consisted of Boehringer Ingelheim, Schering A.G. in Germany, and Parke-Davis's neighbor and rival, Upjohn in Kalamazoo, Michigan, 100 miles west of Ann Arbor.

By the time that Parke-Davis sent out feelers in early 1996 about their intention of finding a comarketing partner for Lipitor, the drug's spectacular efficacy was already known in the industry. In fact, after having heard a presentation on Lipitor's efficacy at a conference in Montreal in 1994, a Merck scientist suggested that it should be called "Turbostatin." Interest level was therefore very high for copromoting Lipitor. Many companies went to Warner-Lambert headquarters in Morris Plains, New Jersey, to negotiate. After the 1971 merger between Warner-Lambert and Parke-Davis, the research and development headquarters remained in Ann Arbor, whereas Morris Plains remained as the business headquarters. Among many suitors, Pfizer Inc. was finally selected by Warner-Lambert as the comarketing partner in the fall of 1996, a few months before Lipitor's approval by the FDA. Lawyers from Warner-Lambert and Pfizer hashed together a copromotion agreement. Within $1 billion of Lipitor's annual sales, a greater portion of the sales would go to the originator, Warner-Lambert. Any sales that exceeded $1 billion would be split equally between the originator and the copromotion partner, Pfizer. The lawyers from Warner-Lambert felt they negotiated the deal of their lives—few believed that Lipitor sales would exceed $1 billion even at its peak.

On the evening of December 17, 1996, Parke-Davis held its annual company Christmas party. The CEO of Warner-Lambert, Lodewijk J. R. de Vink, flew from Morris Plains to Ann Arbor to celebrate the holidays with his research and development colleagues. A few minutes before 8 P.M., as the party seemed to be drawing to an end, Byron Scott, director of worldwide regulatory affairs, rushed in with a fax in his hand.

At 7:40 P.M., the FDA had sent Scott a fax in reference to NDA 20–702:

> Dear Mr. Scott:
> ...We have completed the review of this application, including the submitted draft labeling, and have concluded that adequate information has been presented to demonstrate that the drug product is safe and effective for use as recommended in the draft labeling. Accordingly, the application is approved effective on the date of this letter.

Needless to say, the company Christmas party did not end at 8 P.M. as initially planned.

CHAPTER 6

To Market, to Market

Parke-Davis submitted Lipitor's New Drug Application to the FDA in June 1996 and received approval in December 1996, a relatively short turnaround due to its priority review status. At the time, four statins were already on the market: Merck's lovastatin (Mevacor, 1987) and simvastatin (Zocor, 1991), Bristol-Myers Squibb's pravastatin (Pravachol, 1991), and Sandoz's fluvastatin (Lescol, 1994). The sixth statin, Bayer's cerivastatin (trade names Baycol and Lipobay), another optically pure synthetic statin like Lipitor, was soon to be on the market. The success of the four existing statins had already educated physicians about the benefits of lowering cholesterol levels. More important, Merck's "4S" clinical trials (the Scandinavian Simvastatin Survival Study; see chapter 3) decisively demonstrated the positive impact of lower cholesterol levels in decreasing coronary heart disease. As a consequence, not only did the FDA approve Lipitor in only six months, but it was also widely and warmly accepted by general practitioners and patients immediately after it was available.

When the drug was launched at the beginning of 1997, Warner-Lambert chose Pfizer as its comarketing partner because Pfizer had the strongest sales muscle in the drug industry. Today it is hard to imagine that when Pfizer was founded in 1849, it was a modest laboratory in Brooklyn, New York. By 2008, more than 150 years later, it had grown to become the largest pharmaceutical company in the world.

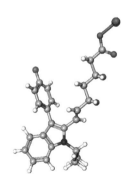

Figure 6.1 Molecular structure of Lescol.

Pfizer Inc.

The small drug firm in Brooklyn made big

In the revolutionary year of 1848, thousands of Europeans immigrated to America to seek new opportunities. Among them were a chemist (the equivalent of today's pharmacist), 20-year-old Charles Pfizer, and his brother-in-law, 22-year-old confectioner Charles Erhart, from the small town of Ludwigsburg in Wuerttemberg, Germany.[1] Unlike many German immigrants at the time who immediately joined the Gold Rush, Pfizer and Erhart decided to stay in New York City and make a living by taking advantage of the crafts that they learned in Germany. With $2,500 borrowed from Pfizer's father and a $1,000 mortgage,[2] they bought a small brick factory in the Williamsburg section of Brooklyn, largely a German neighborhood. Thus, Chas. Pfizer Co., Inc., Specialists in Fine Chemicals was founded. Because most of its employees were German immigrants, New Yorkers nicknamed the company the "German Navy Yard" because it was adjacent to the Brooklyn Navy Yard.

Since its start in 1849, Pfizer had a knack for making better medicines by improving upon existing drugs. They made their first breakthrough when they took a bitter treatment for parasitic worms, blended it with almond-toffee flavoring, and shaped it into a candy cone called Santonin. The company soon established a reputation for producing high-quality chemicals. In 1880, Pfizer began using concentrated lemon juice to make citric acid. At the time, citric acid was the most widely used organic acid in the food and beverage field, but it was produced chiefly in Europe. In 1914, Pfizer established a research laboratory to find a method of making citric acid from sucrose. Five years elapsed before the scientists had successfully "trained" a common black mold, *Aspergillus niger*, found on stale black bread, to transform ordinary sugar into citric acid in pilot plants. In 1923, the company began producing citric acid from the vegetative fermentation of sugar and broke the foreign monopoly with its new process. Curiously, Pfizer decided not to patent the process, but instead swore everyone who knew the production method to secrecy. Because of the insatiable demand for citric acid, the company had to work around the clock, 365 days per year to produce large enough supplies.[3] In 1934, another major breakthrough came when Pfizer developed a method of citric acid preparation by using molasses instead of the much more expensive sugar, saving millions of dollars in raw material and production costs. After that, Pfizer began producing other chemicals by fermentation, including oxalic acid, gluconic acid, and sorbose.

Despite those initial discoveries, Pfizer did not catch most people's attention until its stellar contributions to the mass production of penicillin.

Figure 6.2 The penicillin mold © Royal Mail.

Penicillin

Penicillin, a true wonder drug, has saved million of lives. In fact, without penicillin, 75% of us would probably not be alive and reading this book at this very moment because some of our parents or grandparents would have succumbed to infections. Alexander Fleming discovered penicillin in 1928 in London, England. Seven years passed before Howard Florey at Oxford University took notice. Along with biochemist Ernst Chain, Florey isolated enough penicillin to test it on animals and humans, with great success. Florey's initial fermentation process developed at Oxford had a yield of 0.0001%—about the amount of gold one finds in seawater. However, the raging Battle of Britain prevented them from further investigating and manufacturing penicillin. In order to supply the Allied troops with enough of the drug, a more efficient mass production process was desperately needed. Florey turned to America for help.

The U.S. War Production Board and private pharmaceutical companies took on the challenge. Because of Pfizer's expertise in fermentation accrued from its citric acid production, it was not surprising that Pfizer was one of the companies chosen when the U.S. government launched a top-secret program in 1943 to produce penicillin in large quantities.[4] The "Big Three," Charles Pfizer & Co., Merck & Co., and E. R. Squibb & Sons, were the first companies to participate. Pfizer in particular, then more a chemical company than a bona fide pharmaceutical company, made its name through the penicillin endeavor. Pfizer President George A. Anderson risked $3 million of Pfizer's cash to build a new plant with 14 ten-thousand-gallon fermentation tanks.[2] Using a "deep-tank fermentation" process involving sterilized air continually pumped through the tank, Pfizer became the first to mass-produce penicillin. In the end, the company produced 90% of the penicillin that went ashore with the Allied Forces on the beaches of Normandy in June 1944. Regrettably, because it did not have the name recognition of a drug maker, and without a sales force, Pfizer sold penicillin in bulk to the more established and prestigious drug firms, including Lilly, Parke-Davis, and Upjohn, each of which then distributed the product under their own labels. This humiliating experience served as a rude awakening. When Pfizer discovered its first drug, oxytetracycline (Terramycin) in 1950, the company hired its own sales force and marketed the drug itself.

Drugs of its own

Oxytetracycline (Terramycin) belongs to a class of broad-spectrum antibiotics collectively called the tetracyclines. Aureomycin, the first antibiotic in the tetracycline class, was discovered by 73-year-old botanist Benjamin M. Duggar in 1945. After retiring as a professor of botany, Duggar served as a consultant to Lederle Laboratories at Pearl River, New York, a subsidiary of the American Cyanamid Company. Lederle screened myriad soil samples in search of antibiotics that possessed a better safety profile than streptomycin for treating tuberculosis. A sample from the University of Missouri, where Duggar taught botany 40 years earlier, yielded a very efficacious yellow antibiotic later named chlortetracycline. In 1948, Lederle marketed it under the trade name Aureomycin because of its golden color (*aurum* is Latin for gold). Thanks to its great oral bioavailability, Aureomycin won a good share of the antibiotics market. Nowadays, Benjamin Duggar is considered the pioneer of the tetracycline antibiotics.

As the price of penicillin plummeted, Pfizer became increasingly concerned about competitors' resurgence in the antibiotics arena. Like many pharmaceutical companies at that time, Pfizer plunged into research for newer antibiotics. Every imaginable means was exploited for soil-sample collecting. Travelers, missionaries, explorers, airline pilots, students, housewives, and Pfizer sales agents were encouraged to pick up a teaspoon of earth, seal it in a packet, and mail it back to the company for a small reward. Soil samples rushed in from the most unlikely places: the jungle of

Brazil, the top of mountains, the bottom of mine shafts, cemeteries, deserts, and even the ocean. Balloons were even sent up to collect soil that was airborne. More than 100,000 soil samples were screened, and 75 possible antibiotics were identified. But none made it all the way to the market due to problems with either efficacy or safety—except Terramycin.

In 1949, a yellow powder with strong antibiotic properties was isolated from a soil sample and was subsequently labeled PA-76 (PA stands for Pfizer antibiotic).[2] This sample provided a broad-spectrum antibiotic that proved to be both safe and effective against a range of bacteria responsible for more than 100 infectious diseases. Because it was the 76th antibiotic that Pfizer chemists isolated, they promptly dubbed it the "Spirit of '76," with a nod to the 1776 American Revolutionary War. The soil organism was *Streptomyces rimosus*, the same genus of fungi responsible for producing streptomycin and Aureomycin. The compound was generically known as oxytetracycline. Backtracking revealed that the soil sample, surprisingly, had been obtained from the ground at Pfizer's factory in Terre Haute, Indiana. Oxytetracycline was sold under the trade name Terramycin. The suffix "-mycin" signifies that the drug originated from a mold.

Terramycin was the first drug discovered solely by Pfizer itself, while it was still smarting from its painful experience with penicillin. In 1950, the company established its own sales force, consisting of a grand total of eight salesmen, and sold Terramycin itself. When the drug was approved by the FDA that year, the eight Pfizer sales representatives were already deployed and waiting for the word at pay phones across the nation. They also initiated an aggressive advertising campaign in medical journals, which was controversial at the time because the $8 million in advertising was twice the cost of Terramycin's discovery and development. William C. Steere, Jr., the company's chairman and chief executive officer from 1991 through 2001, began his career as a Terramycin sales representative. After Terramycin hit the market in March 1950, it went on to become a top seller, and Pfizer stock value subsequently doubled in one year.

In addition to its production efforts, Pfizer formed a team to elucidate the chemical structure of oxytetracycline. They enlisted the help of Robert Burns Woodward at Harvard. In 1952, Pfizer and Woodward jointly published their elegant work on the structure of oxytetracycline in the prestigious *Journal of the American Chemical Society*. Meanwhile, a member of the team, Lloyd H. Conover, shocked his colleagues by chemically preparing another powerful antibiotic from chlortetracycline. Under carefully

Figure 6.3 Lloyd Conover (right, with the tie) discusses his research with a colleague © Pfizer.

controlled conditions, using hydrogen gas and catalytic palladium on charcoal, Conover converted Lederle's chlortetracycline to tetracycline by stripping the chlorine atom and replacing it with a hydrogen atom.[5]

Conover's feat was truly revolutionary, because until that time it was generally believed that "natural" antibiotics produced by microbial metabolism were the only ones possessing desirable biologic properties. Conover demonstrated that chemical manipulations could produce active antibiotics, as well. Pfizer branded the new compound Tetracycline. Compared with both Aureomycin and Terramycin, Tetracycline was less toxic. It penetrated the cerebrospinal fluid to a much better extent and was readily absorbed from the gastrointestinal tract and bloodstream. Thanks to these favorable attributes, Tetracycline became the most prescribed broad-spectrum antibiotic in the United States within three years. Pfizer's Tetracycline and Lederle's Aureomycin grabbed 92% of the broad-spectrum antibiotics market. Furthermore, Conover's discovery created a brand-new field of medical research: semisynthetic antibiotics. It sparked a wide-scale search for superior structurally modified antibiotics, which have provided most of the important antibiotic discoveries since then. For his discovery, Lloyd Conover was inducted into the American Hall of Fame for Inventors—only 98 people have earned this honor, including Thomas Edison and the Wright Brothers.

Ironically, Tetracycline was at the center of a long and contentious patent dispute.[6] The settlement in 1962 merely marked the beginning of a trend of litigation involving important drugs, with Lipitor as the latest and the most conspicuous. In October 1952, Pfizer applied for a U.S. patent on Tetracycline. Three months later, American Cyanamid applied for a U.S. patent on the same drug. The two companies agreed in advance that whoever obtained the Tetracycline patent would grant a license to the loser. In January 1955, the U.S. Patent and Trademark Office granted a patent to Lloyd H. Conover of Pfizer for the manufacture of Tetracycline from chlortetracycline. One day after the patent was issued, Pfizer sued Bristol

Laboratories, E. R. Squibb & Son, and the Upjohn Co. for $50 million, charging infringement of the patent. Robert Porter, general counsel and secretary for Pfizer, even hired private investigator John Broady to find out how the secret formula of Tetracycline had leaked to competitors. Porter paid Broady $60,000 to shadow 50 of Pfizer's employees. Broady also tapped the telephones of Squibb and Bristol-Myers on his own initiative, but found no leak.[7] Two days after the patent was issued, announcement was made of the settlement of a separate lawsuit by American Cyanamid against Bristol-Myers, which had been granted a worldwide nonexclusive license to market Tetracycline manufactured by a direct fermentation process.

The commercial success of Terramycin and Tetracycline in Europe led to the establishment of Pfizer's pharmaceutical research site in Sandwich, England, in 1957. Sandwich was a World War I port from which troops and matériel had been shipped to France. The research staff initially totaled only six people—five chemists and one pharmacologist. The lone pharmacologist was Harald Reinert, who served in Germany's Sixth Panzer Division, Afrika Korps during World War II. It was remarkable that Pfizer hired him when wartime memories of Nazi Germany were still so fresh in people's minds.[2]

Pfizer began to publicly trade its stock in 1942. Four years later, with $900,000 in cash raised from its stocks, Pfizer bought the surplus 30-acre Submarine Victory Yard in Groton, Connecticut, from the War Assets Administration. This became the site of Pfizer Central Research in 1959. At that time, almost two-thirds of Pfizer's management was under the age of 40, with only 5% older than 50, but Wall Street regarded Pfizer as one of the best-managed companies a stockholder could find. Anyone lucky enough to have invested $24.75 for one share when Pfizer stock was first publicly issued in 1942, as of December 1958, would have had nine shares worth $990 plus $162.90 from dividends and sales of rights.[1]

Windfall: The Era of Blockbusters

With its acquisition of Pharmacia in 2003, Pfizer became the world's number one drug company, three years after its hostile takeover of Warner-Lambert. Pfizer's ascension to the top of the pharmaceutical industry traces back to its windfall of several blockbuster drugs that they put on the market in the late 1980s and early 1990s.

In the 1960s, Sir James W. Black showed the world how to discover drugs by tackling specific molecular targets through his discovery of beta-blockers. He revolutionized drug discovery from hunting to engineering by employing rational drug design to discover novel compounds that nature had not thought of. Black shared the 1988 Nobel Prize in Physiology or Medicine with Gertrude Elion and George Hitchings for their discovery of important principles for drug treatment.

Applying the principles of targeting molecules, the drug industry reaped the fruit of basic science. Many important drugs were discovered by targeting molecules such as enzymes and proteins. On the one hand, the drugs saved lives and improved quality of life; on the other hand, the drug industry also garnered a financial windfall. The era of blockbuster drugs began with the emergence of Tagamet and Zantac in 1978 and 1983, respectively—the two acid blockers were the first blockbuster drugs, with annual sales exceeding $1 billion, which also ushered in the golden age of the drug industry.

In 1990, Pfizer ranked only ninth largest in size in the pantheon of drug companies. Since then, it amassed an unparalleled six blockbuster drugs with annual sales greater than $1 billion: Viagra (sildenafil, for erectile dysfunction), Zoloft (sertraline, for depression), Diflucan (fluconazole, an antifungal), Zithromax (azithromycin, an antibiotic), Zyrtec (cetirizine, for allergy), and Norvasc (amlodipine, for hypertension). Thanks to these blockbuster drugs, Pfizer acquired a notable reputation and enough financial strength to become the partner of choice for copromoting the new drugs of smaller companies. It was no wonder that Pfizer was finally chosen as the copromotion partner for Lipitor, which also sealed the fate of Lipitor's originating company, Parke-Davis/Warner-Lambert.

Viagra

Not every single person on Earth has heard of Pfizer, but everybody has heard of Viagra, its erectile dysfunction drug. Viagra rivals Coca-Cola as one of the most widely known brand names in the world. Ironically, the story of Viagra is another example of serendipity in drug discovery.

The saga began in 1985 when two Pfizer chemists, Simon Campbell and David Roberts, in Sandwich, England, put together a proposal to look for hypertension and angina drugs. They recommended searching for compounds that inhibit enzymes called phosphodiesterases (PDEs). At that time, little was known in the field except zaprinast, which inhibits several isoforms of the PDEs as well as a few other enzymes. Using zaprinast as the starting point, five Pfizer chemists led by Nicholas Terrett created sildenafil citrate in 1989. Sildenafil citrate, later known as Viagra, is a selective inhibitor of PDE5, with weak activities against other PDE enzymes.[8] At the time, both the chemical and biologic evidence supported a shift to treating angina (chest pain) by inhibiting PDE5.

Up to this point, nothing was extraordinary about the story of Viagra; it was proceeding just like any other drug discovery and development program. Things became interesting after the clinical trials began. Pfizer started phase I clinical trials of sildenafil citrate in 1991 on healthy male volunteers with the intention to continue to phase II trials of angina. Regrettably, the drug did not have the full range of properties necessary to advance as a potential treatment for angina, and it was therefore logical to terminate the trial program. However, Dr. Ian H. Osterloh, local team leader for the early sildenafil trials, and his team noticed that some men had experienced an unanticipated side effect in phase I trials—sometimes referred to as "unexpected benefits": the drug improved their erections![9] In 1994, Pfizer initiated a phase II trial involving 12 patients with male erectile dysfunction. Ten of them had clear improvement with their erections, warranting further phase II and eventually phase III trials, which had similar success.

In March 1998, the FDA approved sildenafil citrate (trade name Viagra) for the treatment of male erectile dysfunction, and four million prescriptions were filled within the first six months. Viagra, on sale in April 1998, smashed all records in its first three months. Prescriptions for Viagra from April through June totaled 2.9 million, enough for estimated sales of $259.5 million. Such huge sales were easily a record. For example, when Lipitor was introduced in 1997, it had sales of $12.4 million in its first two months.[10] From 2000 to 2005, Viagra's annual sales were $1.34, $1.52, $1.74, $1.88, $1.68, and $1.64 billion, respectively. Sales peaked in 2003 because two rival drugs came out that year: Bayer's Levitra (vardenafil) and Lilly's Cialis (tadalafil). Levitra has a half-life of about 4 hours, similar to that of Viagra, but Cialis's half-life of 17.5 hours is much longer. Neither Levitra nor Cialis had become a blockbuster by the end of 2006.

Trials of Viagra in females with sexual dysfunction have failed to show consistent benefit, but in another interesting turn of events, evidence began to accumulate during the late 1990s that the drug might actually benefit patients with pulmonary hypertension, a rare but very severe condition that typically affects young women and even children. In people suffering from this disorder, the linings of the arteries in the lungs become thickened and "furred up," and eventually the heart is unable to pump blood through the lungs. In the late 1990s, treatment options were very limited and unsatisfactory. Pfizer therefore initiated a full clinical trial program and found that sildenafil citrate improved the function of the hearts and lungs in these patients and improved their ability to walk and exercise. The FDA eventually approved sildenafil citrate for this indication (under the trade name Revatio) in June 2005, the second indication for which the drug had been approved after priority (accelerated) review.

Viagra's life has now come full circle. It started out as a heart drug but did not seem to have the necessary properties. It then became a great success for the treatment of sexual dysfunction, but ultimately has also become a much needed treatment for a serious cardiovascular condition.

Zoloft

Major depressive disorder is one of the most common psychiatric disorders, with an estimated 12-month prevalence of approximately 13% in women and 8% in men. About 19 million Americans suffer from depression each year. In terms of financial burden, major depressive disorder ranks as the fourth most costly illness in the world, with estimated annual U.S. costs of approximately $43.7 billion.

The first really effective antidepressants on the market were tricyclics, as represented by imipramine (Tofranil). While the tricyclic antidepressants work effectively, their side effects are troublesome. They are especially toxic when overdosed, even producing cardiac problems and seizures in extreme cases. These unwanted side effects limit compliance, with as few as 1 in 17 patients completing a therapeutic dosing regimen.

In 1988, the FDA approved Eli Lilly's fluoxetine (Prozac), the first selective serotonin reuptake inhibitor (SSRI). Prozac revolutionized the treatment of depression. Although SSRIs are generally *not* more efficacious than tricyclic antidepressants and exhibit a marked delay in their onset of action, they are considerably safer than the older antidepressants. Thanks to better safety profiles, SSRIs have been taken by more than 200 million Americans, making them one of the most administered classes of drugs.

Pfizer's sertraline (Zoloft), while not the first SSRI on the market, became the most prescribed SSRI, with $3.36 billion in sales at its peak in 2004. Mike Wallace, the famous 88-year-old journalist at CBS, began to suffer from depression in the mid-1980s, following accusations of libel and a related lawsuit. He has since gone public with his long-standing fight against depression, even testifying before Senate hearings and urging those who suffer from depression to seek medical treatment. According to various news magazine articles, he now takes Zoloft.

In 1977, Pfizer did not even have an official SSRI program. B. Kenneth Koe, a pharmacologist at Pfizer's labs in Groton, Connecticut, had an idea about norepinephrine uptake blockers. He requested Willard Welch to make some compounds to prove his theory. Welch, a senior medicinal chemist who always enjoyed working at the bench, happily obliged. After only half a year on this project, in February 1978 he made a mixture containing sertraline and its inactive enantiomer. A month later, Welch removed the inactive enantiomer by resolution and obtained pure sertraline.[11] Not only was it a very potent SSRI, but it was also orally bioavailable. Most important, it was stunningly safe at approved doses. After a decade of clinical trials, sertraline was finally approved by the FDA for the treatment of depression in 1991, and Pfizer sold it under the trade name Zoloft.

In 2006, the Zoloft team was bestowed the American Chemical Society Award for Team Innovation.[12] Team members and others familiar with their work agreed that the drug, and the team that discovered it, evolved through a serendipitous process of scientific inquiry. Members of the team were organic chemist Reinhard Sarges, 70; biochemist B. Kenneth Koe, 80; organic chemist Willard M. Welch, 61; animal behavioral scientist Albert Weissman, 72; and Pfizer's head of central nervous system drug research, Charles A. Harbert, 65. All were recognized leaders in their fields of science. Zoloft was the first enantiomerically pure SSRI drug to hit the market. It was also the first bona fide single enantiomer developed by Pfizer. From 2000 to 2005, Zoloft's annual sales were $2.14, $2.36, $2.74, $3.11, $3.36, and $3.26 billion, respectively. Zoloft's U.S. patent expired in 2006.

Diflucan

Fungi, which include yeast and molds, are parasitic plants that help organic matter decompose. Many fungi are our friends: mushrooms garnish our plates; baker's yeast is used in making bread and cakes; cheese is made with the help of fungi; penicillin is the product of a green mold; and last but not least, natural statins, including mevastatin, Mevacor, and Pravachol, are all secondary metabolites of fungi.

Unfortunately, not all fungi are so beneficial; some can even cause deadly diseases. Some medicines have the unfortunate side effect of immunosuppression, rendering their takers particularly susceptible to fungal infections. This is true of chemotherapy in general, as well as for the specific agent cyclosporine, an immunosuppressant for transplant patients. More important, AIDS patients are highly susceptible to a host of fungal infections due to the immunosuppressive impact of HIV viral infection.[13]

In 1939, the first modern antifungal emerged with the isolation of griseofulvin by British scientists. Americans came up with amphoterin B and nystatin in the 1950s. Nystatin, not related to cholesterol-lowering statins, was so named because two scientists from New York State were responsible for its discovery. From then on, a profusion of antifungal drugs emerged, including clotrimazole by Bayer A. G. and miconazole (Lotrimin) by Janssen.

Pfizer Central Research in Sandwich began its search for a novel antifungal in the early 1970s. By 1974, using Janssen's miconazole as its starting point, a group led by Dr. Geoff Gymer discovered tioconazole (Trosyl). Both tioconazole and its prototype, miconazole, are rapidly metabolized in the body and therefore not very useful as oral drugs because they must be taken several times a day. In 1978, Pfizer assembled another team, led by Dr. Kenneth Richardson, to continue search for a novel antifungal drug. Right around that time, Janssen also launched ketoconazole (Nizoral), which was a vast improvement over most other antifungal drugs then available.

In early 1981, Richardson made a major change of their molecules by replacing the imidazole ring with a triazole, resulting in a compound that was three times more active than Janssen's ketoconazole. If one triazole was active, two triazoles might be even better, Richardson supposed. His technician, Bill Million, promptly synthesized the compound, which was extremely potent but toxic to the liver. By Christmas 1981, under Richardson's suggestion, Million prepared a drug by replacing the two chlorine atoms on that toxic compound with two fluorine atoms. The resulting drug, later known as fluconazole (Diflucan), was not only astonishingly potent, but also very safe. Moreover, it was metabolically stable, with a half-life of 29 hours in humans, enabling a once-daily dosing regimen. It could be taken either orally or intravenously. At the beginning of 1990, fluconazole was approved by the FDA.

When fluconazole was initially in development, fungal skin infections and vaginal thrush were its only indications. Nobody thought it would ever achieve blockbuster drug status. However, with the advent of aggressive chemotherapy for cancer, leaving patients prone to infection, and with the increasing prevalence of AIDS, more patients were susceptible to fungal infections. Pfizer conducted further trials at much higher doses in these groups of patients and found other important indications for this drug. It slowly became the world's leading antifungal drug.[13,14] From 2000 to 2005, Diflucan's annual sales were $1.01, $1.07, $1.11, $1.18, $0.94, and $0.50 billion, respectively. Pfizer lost its patent exclusivity for Diflucan in the United States in July 2004.

Zithromax

The first macrolide isolated from nature, erythromycin, was an important antibiotic discovered by Eli Lilly's J. M. McGuire. However, since its marketing in 1952, no follow-up had been discovered by 1974, because all improvements made to the molecule in the ensuing two decades had failed. To make matters worse, erythromycin was unstable in stomach acid, limiting its usefulness.

The story of Zithromax goes back to 1974, when Pfizer Central Research in Groton, Connecticut, initiated a research effort to discover a macrolide with improved antibiotic activity. The Pfizer team intended to create a macrolide with improved bioavailability by modifying erythromycin through semisynthesis.

In the ensuing seven years, the team made more than 2,000 compounds, of which 13 advanced into toxicology studies, and 8 into human trials. Yet despite this massive effort, none had blood levels two to four times higher than erythromycin, as desired. In the end, Pfizer management became disenchanted by the team's lack of progress after so many years and was on the verge of closing down the program. But the tide began to turn in early 1981—chemists on the team came across a patent by Pliva Pharmaceuticals, a company in Zagreb, Croatia, then part of Yugoslavia. Pliva was doing essentially what Pfizer was trying to do to improve upon erythromycin. Two Pfizer chemists resynthesized Pliva's drug, but it was not orally active because it was metabolized rapidly in the stomach. However, in May 1981, research chemists G. Michael Bright and Dick Watrous methylated an amine on Plivia's drug, giving rise to a compound that would become azithromycin (Zithromax). The drug exceeded their expectations by being orally active, acid stable, and effective against *Haemophilus influenzae*, and it even stayed in the body tissue of animals longer than other antibiotics. Azithromycin was generated just days before the project's anticipated termination.[2]

Pfizer applied for and was granted a patent for azithromycin by the U.S. Patent and Trademark Office in 1982. Interestingly, Pliva applied for and was granted a patent on the same compound as well a year before. Thus, two companies held U.S. patents on the same compound, due to an oversight on the part of the Patent Office. Because Pliva's Dr. Slobodon Djokic actually discovered azithromycin before Pfizer did, the two companies signed a copromotion agreement. As a result, Pfizer abandoned its patent in return for the rights to market the drug throughout most of the world, while Pliva would sell in the Eastern Bloc countries, including Russia.[2] Zithromax was finally approved by the FDA in 1992.

The common misconception is that "a good drug sells itself." In reality, creative marketing normally plays a significant role, as exemplified by Pfizer's promotion of Zithromax. At the very beginning, Pfizer's marketing message focused on consumers. Because the drug stays in the body for such a long time, patients need fewer doses than they would with other antibiotics. Pfizer's marketers knew that this would be a great selling point for parents struggling to get their sick children to take medicine. The marketing message to pediatricians was "once a day dosing for just five days and you're done."[15] When the pediatric formulation of Zithromax was approved in 1995, some pediatric clinics were decorated with Zithromax zebras, a stuffed animal doll supplied by Pfizer sales representatives. The doll's name was "Max," short for Zithromax, which became a billion-dollar drug in just a few years. It was the nation's top-selling branded antibiotic.[16] From 2000 to 2005, Zithromax's annual sales were $1.38, $1.51, $1.52, $2.01, $1.85, and $2.02 billion, respectively. Zithromax lost its patent protection in 2005.

Norvasc

Norvasc (amlodipine), a calcium channel blocker, lowers blood pressure, which is another risk factor for heart attacks. Calcium channel blockers, also known as calcium channel antagonists or calcium entry blockers, are widely used in the treatment of hypertension, angina, and rapid heartbeat (tachycardia), including atrial fibrillation. When a calcium channel blocker enters the opening of a calcium channel, the drug gets stuck (like a fat man caught halfway through a porthole), thus preventing calcium ions from passing through the channel. As a consequence, calcium channel blockers slow or block the transport of calcium ions into muscle cells in blood vessel walls, thus reducing contraction of blood vessels and lowering blood pressure.

In 1969, Professor Kroneberg, a leading pharmacologist at Bayer A. G., contacted a pioneer in the study of the calcium channels, Professor Albrecht Fleckenstein at the Physiological Institute of the University of Freiburg in Germany. Kroneberg handed Fleckenstein two compounds, Bay-a-1040 and Bay-a-7168. Fleckenstein determined that these two drugs were strong coronary vasodilators and that they indeed selectively blocked calcium channels. Bay-a-1040, with a 1,4-dihydropyridine core structure, was later given the generic name nifedipine. Bayer began marketing it with the trade name Adalat in 1975. Pfizer licensed the marketing rights for nifedipine in the United States from Bayer and sold it under the trade name Procardia. Nifedipine harbingered one of the most important classes of calcium channel blockers: 1,4-dihydropyridines. Unsatisfactorily, nifedipine is a short-acting calcium channel blocker and has to be taken several times a day, necessitating second-generation drugs with improved profiles.

Using Bayer's nifedipine as the prototype, Pfizer Central Research in Sandwich embarked on its own calcium channel blocker program in the early 1980s. Pfizer wanted to improve metabolic stability so that the drug could be taken only once a day. The team prepared hundreds of 1,4-dihydropyridine analogs. In July 1981, Pfizer chemist Nick Smith, working with his supervisor, John Stubbs, synthesized the drug that would later become amlodipine (Norvasc). Amusingly, amlodipine had been previously isolated as a metabolite from dog urine after the dog was given another 1,4-dihydropyridine drug.

In contrast to nifepidine, amlodipine has a higher bioavailability and a longer half-life in plasma, so it can be taken once daily. Such long duration of action makes it ideal for the long-term treatment of hypertension. Moreover, Norvasc *gradually* reduces blood pressure. This is a good attribute because reducing blood pressure too quickly can cause fainting spells, which is a side effect with some of the other 1,4-dihydropyridine calcium channel blockers. All of these features helped make Norvasc the most prescribed antihypertensive agent in the world. Norvasc was approved by the FDA in July of 1992. From 2000 to 2005, Norvasc's annual sales were $3.36, $3.58, $3.85, $4.34, $4.46, and $4.71 billion, respectively, making it Pfizer's second-best selling medicine immediately behind Lipitor.[17]

Within the last several years, generic drug companies have become extremely aggressive in challenging brand-name drugs' patents so that they can make and sell the copycat versions of those drugs. This is especially true for blockbuster drugs. Pfizer's patent on amlodipine (Norvasc) expired on September 25, 2007. However, Dr. Reddy's Laboratories, a generic drug company based in Hyderabad, India, intended to sell a different chemical version of amlodipine five or six years before the patent was to expire. In December 2002, a New Jersey court ruled that Dr. Reddy's Laboratories could sell a generic version of Pfizer's Norvasc. Pfizer appealed the decision, and about a year later, the U.S. Court of Appeals for the federal circuit in Washington, D.C., overturned the New Jersey judge's ruling. The appeals court ruled that Pfizer's patent covers all forms of amlodipine, the crucial ingredient in Norvasc, including the version proposed by Dr. Reddy's, until 2007.[18]

Interestingly, similar patent litigation took place for Lipitor in 2005. Ranbaxy Ltd., another generic drug company in India, challenged the validity of Pfizer's patent on atorvastatin (Lipitor), but Pfizer prevailed again.

The Merger

Success can often be a double-edged sword. Lipitor made billions of dollars for Parke-Davis's parent company, Warner-Lambert, but it also sealed the fate of the company, eventually sending it to the corporate graveyard. The saga of Pfizer's acquisition of Warner-Lambert was one of the most fascinating stories of modern capitalism at its finest.

Thanks primarily to Lipitor's spectacular efficacy and safety, and thanks in part to Pfizer's marketing prowess, Lipitor became a blockbuster drug within its first year on the market in 1997, the first and the only drug to do so. Not surprisingly, Warner-Lambert and its copromotion partner, Pfizer, enjoyed tremendous prosperity from the Lipitor fortune.

In 1999, Warner-Lambert was on a roll. The company's Lipitor and Rezulin (troglitazone), a diabetes drug that was later withdrawn, were both doing very well. Warner-Lambert had also signed a deal with Forest Laboratories to copromote Celexa (citalopram), an SSRI antidepressant.

The company's stock rose sharply in those days, tripling in value in only three years to approximately $130 per share. In November, Warner-Lambert's CEO, Dr. Lodewijk J. R. de Vink, surprised the industry by announcing a $58.3 billion friendly merger of equals with American Home Products. Fearing the loss of its cash cow Lipitor, Pfizer stunned Wall Street by immediately proposing a $72 billion hostile takeover. In response, American Home Products raised its bid to $70 billion. The next day, Pfizer countered with a sweetened offer of $82.4 billion.[19] The bidding war kept heating up in what became one of the drug industry's nastiest takeover battles, and at stake was control of Lipitor. On November 16, 1999, Warner-Lambert released some confidential details of its agreement with Pfizer to copromote Lipitor, which revealed how valuable Lipitor was to Pfizer's future performance. Warner-Lambert also reiterated that it was considering canceling its copromotional agreement with Pfizer altogether. Some analysts questioned whether Pfizer had miscalculated in its attempt to take over Warner-Lambert, especially if the company was to lose its stake in Lipitor revenue altogether.[20]

On November 23, 1999, Pfizer sued Warner-Lambert, contending that the company had breached their joint marketing contract. This lawsuit was part of Pfizer's legal strategy to break up Warner-Lambert's deal to merge with American Home Products. Not to be outdone, in an effort to remain independent, Warner-Lambert countersued Pfizer a week later in an attempt to end the Lipitor marketing agreement. On December 17, Pfizer filed plans seeking to replace the board of Warner-Lambert with directors who did not oppose Pfizer's hostile takeover bid. As if the war was not interesting enough, Procter & Gamble joined the fray in mid-January 2000, playing the "white knight" to save Warner-Lambert from its quandary. Unfortunately, investors reacted violently, sending Procter & Gamble's stock nosediving, with 19% loss of value overnight. The company promptly, and wisely, decided to stick with what it knew in consumer products and backed off from the deal with Warner-Lambert.[21]

In early February, Pfizer increased its bid to $87 billion. Warner-Lambert executives finally capitulated in favor of the proposed merger under tremendous pressure from the shareholders—after all, Pfizer's offer presented a huge premium. After an acrimonious three-month takeover battle and a record-setting $1.8 billion breakup fee paid to American Home Products, Pfizer acquired Warner-Lambert for $90.27 billion in stock. Warner-Lambert's chairman de Vink stepped down. Pfizer then became the world's second largest pharmaceutical company, right behind

GlaxoSmithKline. In June of 2000, the Federal Trade Commission approved the merger.

In the early 1990s, the Warner-Lambert Company was worth around $12 billion. With Lipitor, Warner-Lambert's value jumped to roughly 10 times that amount. The acquisition of Warner-Lambert gave Pfizer the full marketing rights to Lipitor, which they then built into the world's best-selling drug, earning $12.9 billion in sales in 2007.

By consummating this merger, Pfizer gained full ownership of Lipitor. By then, it owned an unparalleled seven blockbuster drugs: Viagra (sildenafil, for erectile dysfunction), Zoloft (sertraline, for depression), Diflucan (fluconazole, an antifungal), Zithromax (azithromycin, an antibiotic), Zyrtec (cetirizine, for allergy), Norvasc (amlodipine, for hypertension), and now Lipitor (atorvastatin, to lower cholesterol)—the crown jewel of all drugs.

Half a year after the Warner-Lambert acquisition, Pfizer CEO William C. Steere, Jr., passed the helm to Henry "Hank" A. McKinnell, who became the CEO at the beginning of 2001. That same year, Pfizer launched its atypical antipsychotic for the treatment of schizophrenia, ziprasidone hydrochloride (Geodon). Unlike many other atypical antipsychotics, Geodon does not have the weight gain adverse effect. In 2002, Pfizer introduced voriconazole (Vfend), an orally and intravenously administered antifungal indicated for the treatment of serious fungal infections.

Because buying Warner-Lambert, and Lipitor with it, was such a boon to Pfizer's bottom line, Hank McKinnell masterminded an encore. In April 2003, Pfizer acquired Pharmacia Corporation for $57 billion. With this merger, Pfizer became the world's largest drug company, overtaking GlaxoSmithKline. With this deal, Pfizer increased its collection of blockbusters with celecoxib (Celebrex), a COX-2 inhibitor for arthritis. The story of Celebrex gives one a strong sense of déjà vu: as with Lipitor, when Celebrex was approved by the FDA in 1999, Pfizer was enlisted by Pharmacia as the copromotion partner. Pfizer ended up gobbling up its partner Pharmacia to gain full control of Celebrex, just like it did with Warner-Lambert to gain full control of Lipitor.

PROVE-IT

For existing drugs on the market, drug companies often carry out phase IV studies in order to demonstrate that they can help patients suffering from other illnesses, or to prove that they work better than competitors' drugs. On the one hand, these trials expand the patient base; on the other hand, they expand the claims that the sales representatives can make about the drugs. After Lipitor's approval for marketing, Pfizer carried out 400 clinical studies involving more than 80,000 patients. The drug is easily one of the best-understood drugs in history.

One phase IV trial bestowed the greatest boon for Lipitor's reputation. Ironically, it was not even conducted by Pfizer. Instead, it was carried out by a rival, Bristol-Myers Squibb, whose Pravachol was on the market in the United States in 1991. In order to improve its third-place position in sales in the statin market behind Zocor and Lipitor, Bristol-Myers Squibb provided funding to initiate a large clinical trial called PROVE-IT, an acronym for pravastatin or atorvastatin evaluation and infection therapy. It was an ambitious trial that involved 4,162 patients from 349 sites in eight countries. The primary purpose of this groundbreaking trial was to gauge the effectiveness of Pravachol and Lipitor at reducing cardiovascular events in patients with an acute coronary syndrome. Another purpose was to see if extremely low LDL-cholesterol levels could achieve a greater reduction in cardiovascular events.

Patients enrolled in the trial had been hospitalized for a sudden attack of chest pain from heart disease. The treatment groups were given either 40 mg of Pravachol or 80 mg of Lipitor, both the highest approved doses at the time (Pravachol was approved for the 80-mg dose in 2004). The results of this two-year study were published in April 2004 in the *Journal of the American Medical Association*.[22] The Lipitor patients achieved an average LDL level of 62 mg/dL, compared with "only" 95 mg/dL for the Pravachol group, a level that was considered very successful prior to the new national guidelines for cholesterol. PROVE-IT thus produced a surprise slam dunk for Lipitor. The patients taking Lipitor were significantly (16%) less likely to have heart attacks or to require bypass surgery or angioplasty within a week or so after being hospitalized for a heart attack or unstable angina.[23] Additionally, within one month, the Lipitor patients fared better than those taking Pravachol in terms of lower rates of heart attack, bypass surgery, angioplasty, or death.[24]

More important, this trial confirmed the mantra of many cardiologists that "lower is better" with regard to LDL-cholesterol levels. Lowering cholesterol levels far beyond the levels recommended by most doctors could substantially reduce heart patients' risk of suffering or death due to a heart attack. The trial also confirmed that reducing cholesterol levels far below 100 mg/dL could be beneficial. Gina Kolata, a science reporter from the *New York Times*, interviewed prominent cardiologists from across the country about their response to the PROVE-IT trial.[23] Dr. Christie M. Ballantyne, a professor at Baylor College of Medicine, commented: "It looks as if the study's name—PROVE-IT—was entirely apt, if not in the way intended. They did prove it, but I don't think they proved what they thought they would prove. It is remarkable." Dr. Steven E. Nissen of the Cleveland Clinic said that drug companies "never, ever sponsor a trial like this that they think has a chance of going the wrong way. This trial backfired because in fact the differences between these two drugs are very profound." Researchers said they were particularly surprised because the study was intended to show that Pravachol was just as effective as Lipitor. Dr. Andrew G. Bodnar of Bristol-Myers Squibb stated that it was hard to imagine that simply lowering LDL-cholesterol levels could make a significant difference within a month. One possibility, he suggested, was that more intensive therapy had a greater effect in suppressing inflammation. When a plaque is inflamed, it was more likely to burst open.[24]

Dr. Bodnar seemed to be in good company with his scientific insight in interpreting data from PROVE-IT. Half a year after PROVE-IT results were out, disappointing results of the "A to Z Trial" that tested high doses of Zocor were published in the *Journal of the American Medical Association*. In this 4,500-patient study, patients taking high doses of Zocor fared no better than a low-dose group in terms of risk of heart attack, despite lower LDL-cholesterol levels. Why? Perhaps because Lipitor works not only by lowering cholesterol but also by reducing inflammation, something Zocor cannot do as well, experts speculated.[25]

No fewer than 400 phase IV clinical trials have been carried out by Pfizer for Lipitor, each with a catchy acronym, for example, ALLIANCE, ASCOT, AVERT, CARDS, CARE, CURVE, IDEAL, MIRACL, REVERSAL, SPARCL, TNT, and WATCH. These trials consistently demonstrated the safety and efficacy of Lipitor in both lowering the LDL-cholesterol levels and lowering the heart attack risk. They also repeatedly broadened Lipitor's label of approved uses, changes that must be approved by the FDA. From its marketing at the beginning of 1997 to the end of 2006, Lipitor became far and away the best-selling drug in history. Between 1997 and 2007, its annual sales were as follows: 1997, $1 billion; 1998, $2.2 billion; 1999, $2.7 billion; 2000, $5.03 billion; 2001, $6.45 billion; 2002,

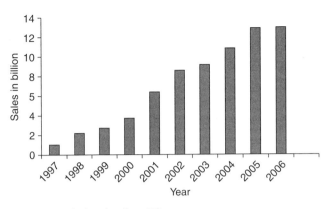

Figure 6.4 Lipitor annual sales, data from Pfizer.

$7.97 billion; 2003, $9.23 billion; 2004, $10.9 billion; 2005, $12.2 billion; 2006, $12.9 billion; and 2007, $12.9 billion.

By now, many landmark trials have shown that long-term therapy with statin drugs reduces the risk of death, myocardial infarction, and stroke among patients with established coronary artery disease even when LDL-cholesterol levels are not elevated. They support a strategy of aggressive LDL-cholesterol lowering following acute coronary syndrome in order to prevent death and other major cardiovascular events. Again, the mantra is still "the lower, the better."

In 2004, Pfizer started selling a combination pill of amlodipine (Norvasc) and atorvastatin calcium (Lipitor), with the trade name of Caduet, the first medicine to treat these two different conditions with one pill. It lowers both blood pressure (Norvasc) and cholesterol (Lipitor) at the same time, allowing physicians to help patients reduce their risk of developing cardiovascular disease with better compliance.

The Patent Litigation

The world has had brand-name drug companies and generic drug companies for more than a century. The old rule of engagement was that generic companies waited until the patents expired and then began marketing the generic drugs at lower prices. That way, brand companies had more than 10 years of exclusive market and could recoup the huge expenses involved in discovering innovative drugs. Recently, generic drug companies have lost their patience in the face of the enormous profit that blockbuster drugs promise and are changing the rules of engagement. The potential profit lures them to legally challenge the validity of the patents for a lucrative brand-name drug many years before they expire. Because the stakes are so high, legal fees are dwarfed by the potential profit. If the challengers are successful, they are given six months of exclusivity for the generic version of the drug, after which all generic companies are then allowed to produce and market the copycat version of the drug, at which point the price inevitably plummets.

A patent is a contract between an inventor and the government, in which the inventor discloses the discovery so others can learn from it. Meanwhile, the government gives the inventor the exclusive right to prevent others from making, using, or selling its invention for a certain period of time. Prescription drugs are protected by patents that give their owners the right to exclude others for up to 20 years, although they usually must spend at least part of that time gaining federal approval. This patent protection enables the drug maker that discovered the drug to earn back its development costs and make a profit. Otherwise, competitors could immediately make and sell identical versions of the medicine, undercutting the company that invented it. If such scenario went unchecked, it could lead to the collapse of the pharmaceutical industry, and new drugs would cease to be invented.

It was calculated that if a generic company could secure the six-month exclusivity period for making the generic version of Lipitor, atorvastatin calcium, the company stood to gain as much as $1 billion. The first challenger was Ranbaxy Laboratories Ltd. in New Delhi, the largest drug company in India, whose size was dwarfed in comparison to Pfizer. In August 2003, Ranbaxy challenged Lipitor patents in the United Kingdom, United States, Austria, Norway, Romania, and Peru. It had hoped to market Lipitor in the United States as early as 2008! That was actually not the first legal clash between the two companies. A few years before, the two companies also crossed swords when Ranbaxy challenged Pfizer's patent extension of Diflucan. In 2004, the U.S. Court of Appeals for the District of Columbia upheld an earlier decision by a lower federal court. The decision acknowledged that Pfizer, because it had completed pediatric trials of the medicine, was entitled to keep the monopoly on Diflucan six months beyond the expiration of its patent.

For Lipitor, two key patents were involved. The basic Lipitor patent (generic invention, U.S. Patent 4,681,893—the "893 patent") covers the active ingredient. The second patent (selection invention, U.S. Patent 5,273,995—the "995 patent") covers the optically pure form (enantiomer) of the active ingredient, atorvastatin. Ranbaxy claimed that its generic form of atorvastatin would not infringe the basic Lipitor patent. They also claimed that the pure enantiomer patent was invalid as well due to "lack of novelty" and "lack of an inventive step." If Ranbaxy were successful, such a victory would allow it to launch a generic version of atorvastatin calcium at a price as much as 50% lower than the price of the branded product Lipitor, several years before the normal patent expiration. Simply because of the possible huge payback, Ranbaxy appropriated $20 million for legal fees for Lipitor litigation alone, which was 2% of its sales and an unheard of ratio in corporate India.

In October 2005, the British judgment was handed down, ruling that a Ranbaxy generic would indeed infringe on Pfizer's basic Lipitor patent. The court ruled that the pure enantiomer patent was invalid. Ranbaxy appealed this decision and in June 2006 the U.K. Court of Appeals upheld the lower court decision. This ruling meant that Ranbaxy would have to wait until June 28, 2011, to release its generic version in the U.K. market. According to news reports, when the British judgment was handed down, Pfizer CEO Hank McKinnell stated: "This is an important victory, not only for Pfizer, but also for all innovators pursuing the high-risk medical discoveries that benefit current and future generations of patients around the world."

In the United States, Ranbaxy and Pfizer met in November 2004 at the federal district court in Delaware (the State of Delaware has favorable corporate laws, so numerous companies have been incorporated there). Both sides hired the best litigation lawyers and called many expert witnesses. Bruce Roth, Roger Newton, and other key players involved with Lipitor's discovery and development were deposed and cross-examined by lawyers from both sides. The trial was the culmination of a bitter and contentious battle that had lasted for several years, leaving Pfizer's shareholders biting their nails because the stakes were so high.

In December 2005, Judge Joseph J. Farnan, Jr., of the Federal District Court of Delaware ruled that Ranbaxy's product infringed on *both* patents that Pfizer held for Lipitor.[26] Ranbaxy appealed this decision to the Federal Circuit Court of Appeals and received a split decision. In August 2006, the U.S. Court of Appeals for the Federal Circuit in Washington, D.C., held that Ranbaxy's generic product atorvastatin infringed on the basic Lipitor patent, which expires in March 2010, but invalidated one claim of the pure enantiomer patent, overturning Delaware Judge Farnan's ruling in that regard. Ironically, the claim of the pure enantiomer patent was found to be invalid for a mere technicality in the way it was worded. Pfizer contended that the ruling was based on a "technical defect" that the company would seek to change. The company might also ask the court to reconsider that aspect of the ruling.[27] This could mean that the generic version of Ranbaxy's atorvastatin could hit the U.S. market by March 24, 2010. It now had the opportunity to bring the launch date forward to March 2010 from June 2011 with 180-day exclusivity in the U.S. market.[28] The loss of 14 months of patent protection for Lipitor would be 14 months of lost revenue worth more than $15 billion!

The saga ended on June 19, 2008, when Pfizer struck a deal with Ranbaxy to keep the generic version of Lipitor off the U.S. market until November 2011. In return, the pact both provides a firm launch date and ensures that Ranbaxy will be the only one eligible to sell a generic version for 180 days after Lipitor's patent expires. In addition, Ranbaxy also secured the right to sell a copycat of Caduet—another subject of litigation between the companies—seven years before that drug's 2018 patent expiration. Caduet, a pill that combines Lipitor and the blood-pressure medicine Norvasc and had $568 million in 2007 sales, has been a disappointment for Pfizer.

CHAPTER 7

Baycol, Crestor, and Cholesterol Drugs beyond Statins

In 2001, the National Heart, Lung, and Blood Institute, a division of the National Institutes of Health, released new guidelines for the treatment of hypercholesterolemia. These guidelines called for patients with an average risk of coronary heart disease to reduce their LDL-cholesterol levels to less than 130 mg/dL; for those at high risk, the target is less than 100 mg/dL; and for the highest risk patients, the target is less than 70 mg/dL. The new guidelines nearly tripled the number of Americans who stood to benefit from drug therapy to get their lower cholesterol levels—to 36 million from 13 million. Statins had already become the drug of choice for cholesterol control. Statin awareness was further raised in the wake of former President Bill Clinton's well-publicized heart attack in September 2004 after he stopped taking Zocor. Increasingly, new data demonstrated the benefits of reducing cholesterol levels even further than physicians had been advocating. The sales of statins therefore skyrocketed, reaching $21 billion in 2004 and $22 billion in 2005. Just like everything else, not all statins are created equal, a point clearly demonstrated by the story of Baycol.

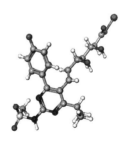

Figure 7.1 Molecular structure of Crestor.

The Story of Baycol

No medication, not even a nonprescription drug like aspirin or acetaminophen that you buy over-the-counter, is 100% safe. Whether or not you realize it, whenever you take a drug, you are weighing the potential benefits against the possibility that the medicine can hurt you. This analysis is what the experts call the *risk–benefit profile* of a drug.

Four months after Lipitor was launched for sale, the FDA approved cerivastatin, the sixth statin, for use in lowering cholesterol in 1997. Germany's Bayer A. G. sold it under the trade name Baycol and Lipobay (for the free acid form) until it was forced to withdraw it from the market in August 2001.

Baycol was discovered by a group of Bayer scientists led by medicinal chemist Rolf Angerbauer and biologist Hilmar Bischoff in Wuppertal, Germany, in the late 1980s.[1] It was only the second optically pure synthetic statin on the market (Lipitor was the first). Baycol was extremely potent, approximately 100-fold more potent than Mevacor, the grandfather of all statins. During initial clinical trials involving approximately 3,000 patients, no serious adverse events were observed, although initially only 0.2-mg and 0.3-mg doses were approved. The 0.4-mg and 0.8-mg doses were also approved a couple years later.[2] With GlaxoSmithKline as its copromotion partner, Bayer pushed Baycol's sales to $554 million in 2000, making it one of the company's fastest-growing products.[3]

All statins are known to cause rhabdomyolysis, a disorder in which muscle cells break down, overwhelming the kidneys with cellular waste and leading to kidney failure. For five statins before Baycol (Mevacor, Zocor, Pravachol, Lescol, and Lipitor), rhabdomyolysis occurs very rarely and is almost never fatal. With Baycol, however, reports of serious rhabdomyolysis were about 15–60 times as frequent as with the other statins. It has been speculated that Baycol was more lipophilic than most other statins and highly bioavailable—its bioavailability was twice of that of Lipitor (60% vs. 30%). Therefore, the drug was distributed all over the body, without tissue selectivity. It is a case of "too much of a good thing gone bad." Bayer priced Baycol so competitively that some overconfident clinicians initially overlooked the higher rhabdomyolysis rate. Its prescriptions soared. Sadly, as the patient population grew exponentially, so did the number serious adverse effects.

A Baycol fatality tended to occur in patients taking higher doses of the drug and in those also taking fibrates, especially gemfibrozil (Lopid),

Parke-Davis's popular fibrate drug that lowers blood triglycerides. Recent studies have shown that the risks of drug–drug interactions increase 1,400 times if statins are combined with fibrates—a textbook example of a drug–drug interaction. Therefore, statins should not be taken with fibrates, since both classes of drugs are metabolized by the same enzyme, CYP 3A4, in the liver. When the liver is busy metabolizing one drug, it then does not have more capacity to metabolize the other, resulting in serious and sometimes fatal consequences. One example is the grapefruit juice effect: many drugs such as statins have drug–drug interaction with grapefruit juice, which should be avoided when taking statins.

Once the risk of coadministration was revealed in January 1999, Bayer immediately added a warning to Baycol's labeling stating that the drug was associated with rhabdomyolysis. A drug's labeling, approved by the FDA, includes a document called a package insert, which is included with each filled prescription. The package insert often spans many pages, and doctors rely on these inserts when they prescribe medicines. In April 1999, it was reported in an article in the *American Journal of Cardiology* that a patient fell seriously ill after taking Baycol with Lopid.[4] Reports of severe interaction between Baycol and Lopid began to appear in the literature.[5] By the end of 1999, Bayer added a stronger warning to the drug's label specifying that Baycol should not be prescribed with Lopid—a warning stronger than those on the labels of similar statin drugs. Meanwhile, Bayer also sent every physician a recommendation not to prescribe Baycol for patients who also took Lopid. Unfortunately, most busy physicians did not heed the label's warning or the letter.

In April 2000, in the *Annals of Internal Medicine*, three Spanish doctors reported that a woman developed rhabdomyolysis after taking Baycol even though she was not taking Lopid.[6] A few months later, six Turkish doctors reported in the journal *Angiology* that a woman died after taking Baycol with Lopid.[7] In July and August 2001, the FDA raised serious concerns about Baycol, and Bayer voluntarily recalled Baycol. By then, at least six million people worldwide had taken the drug, including 700,000 Americans. Ultimately, more than 100 deaths were attributed to Baycol therapy, and many lawsuits ensued.[8] By 2003, Bayer's share price had plummeted 26%.[9] The German pharmaceutical titan, responsible for the discovery of aspirin and Cipro (ciprofloxacin, an antibiotic), was dealt a significant blow by the effects of Baycol. The downfall of Baycol contributed in a major way to the closure of Bayer's West Haven, Connecticut, site, its research and development headquarters in the United States.

Crestor

Rosuvastatin (Crestor), introduced in 2003, was the seventh statin on the market. AstraZeneca, an Anglo-Swedish company, licensed it from a Japanese company, Shionogi & Co. Ltd. Data from comparative trials has confirmed that on a per-weight basis, Crestor, billed as the superstatin, is the most efficacious statin for lowering LDL cholesterol, with reductions of up to 63% reported with a 40 mg/day dose.[10] In clinical trials involving 12,400 patients who received doses of 5–40 mg of rosuvastatin, no serious adverse effects were detected.

Crestor had a bumpy ride even before it was on the market. AstraZeneca initially also tested the 80-mg dose. But in July 2002, the company asked doctors taking part in the clinical trials to stop prescribing the highest dose of Crestor after two patients taking the 80-mg dose in studies suffered kidney damage.[11] AstraZeneca sent letters asking doctors participating in research to switch patients from 80 mg to a lower dose. The FDA had actually delayed its decision on the drug for a year while assessing concerns about its risk to the muscles and kidneys, especially for the 80-mg dose. In August 2003, the FDA approved Crestor in doses of 5–40 mg.[12]

A few months after Crestor was on the market in Europe, an unusual editorial titled "The Statin Wars: Why AstraZeneca Must Retreat" was published in the prestigious medical journal *Lancet*. The journal's editor, Richard Horton, scathingly criticized AstraZeneca's CEO, Tom McKillop, for marketing Crestor despite the lack of outcome trials and still uncertain safety data. Horton's editorial concluded with this statement: "AstraZeneca has pushed its market machine too hard and too fast. It is time for McKillop to desist from this unprincipled campaign."[13] Many observers were baffled because personal attacks on an individual or a product are extremely rare in prominent medical journals. McKillop countered with Crestor's extensive safety data and concluded: "Regulators, doctors, and patients as well as AstraZeneca have been poorly served by your flawed and incorrect editorial. I deplore the fact that a respected scientific journal such as *Lancet* should make such an outrageous critique of a serious, well studied, and important medicine."[14]

A year later, in 2004, Crestor weathered a similar barrage from the drug industry's longtime critic, Public Citizen, a Washington-based consumer advocacy group founded by Ralph Nader. Sidney M. Wolfe, a combative director at Public Citizen who had strongly opposed Crestor's approval since the very beginning, petitioned the FDA to withdraw it from the market because of increased risk for proteinuria (spillage of protein in the urine) and hematuria (spillage of iron in the urine) in addition to rhabdomyolysis.[15] AstraZeneca reasserted that the drug was safe and dismissed the accusations made by Public Citizen as groundless: "There are more than two million patients on Crestor, and we've not seen anything to indicate that the safety profile is any different from the other marketed statins," said Gary Bruell, an AstraZeneca spokesman.[16] After an extensive investigation, the FDA stood by its conclusion with regard to Crestor's safety profile and rejected Public Citizen's petition to remove it from the market.

In another unusual move, appearing before a Senate committee, Dr. David Graham, an FDA safety officer, testified that there were five drugs on the market that he had serious concerns about.[17] Those five drugs were Accutane (by Roche, for acne), Bextra (by Pfizer, for arthritis), Crestor (by AstraZeneca, for cholesterol), Meridia (by Abbott, for obesity), and Serevent (by GlaxoSmithKline, for asthma). Graham cited reports that some patients on Crestor, one of the newest anticholesterol statins, suffered kidney failure. The drug's manufacturer, AstraZeneca, said the prescription drug is safe as long as it is used properly. Indeed, there is little evidence suggesting that Crestor is not as safe as other statins. But because it arrived just two years after Baycol's withdrawal, Crestor was the innocent victim of negative public perception by both physicians and patients.

Although AstraZeneca vehemently defended the safety of Crestor, sales of the drug fell after Public Citizen's petition and Dr. Graham's testimony. Sales finally began to rebound after the company engaged in a marketing campaign to promote the drug as the most potent statin available. Its worldwide sales totaled $908 million in 2004.

In 2005, a controversial article published in the journal *Circulation* revealed mixed results for Crestor. Dr. Richard H. Karas and his team at Tufts University conducted a so-called postmarketing study, in which the rates of adverse reactions were calculated based on reports that doctors voluntarily submitted to the FDA.[18] The team looked for four main types of side effects: rhabdomyolysis, proteinuria, a reduced ability of the kidneys to filter toxic substances from the blood, and kidney failure. Patients taking Crestor were eight times more likely to develop rhabdomyolysis, kidney failure, or proteinuria than patients taking Pravachol, 6.5 times more likely than those taking Lipitor, and 2.2 times more likely than those taking Zocor. Crestor had less than half the rate of adverse events reported for

Baycol, which was removed from the market in 2001. For the four drugs studied, the absolute rate of adverse events per million prescriptions was relatively low: 28 for Crestor, 13 for Zocor, 3.5 for Pravachol, and 4.3 for Lipitor.[18] Overall, this study showed that these statins were safe. After an exhaustive review of the same information used by Dr. Karas's team and Public Citizen, the FDA concurred that all available evidence indicated Crestor had no greater risk of muscle toxicity or serious kidney damage than Lipitor, Pravachol, or Zocor.

Overcoming the controversy over its safety, Crestor sales recovered and totaled $1.27 billion in 2005—respectable, but not as much as the initial prediction of $3–4 billion a year. It controlled 7.6% of the statin prescription market, whereas Lipitor had 55%. Crestor sales were $1.08 billion in 2006, ranking 37th in pharmaceutical retail dollars that year.

In March 2006, Crestor received a healthy boost: the landmark ASTEROID clinical trial showed that Crestor not only halted the progression of heart disease, but also partially reversed it. It could reverse the buildup of plaque in coronary arteries, giving it a competitive edge over other statins.[19] The study, involving 507 patients, provided each with Crestor at a dosage of 40 mg/day, the highest approved dose for the drug. After two years, of the 349 patients whose plaque volume could be evaluated, Crestor had reduced their arterial deposits by between 7% and 9%, an unprecedented feat. A previous trial found that Lipitor had demonstrated only stabilization in plaque volumes, although it reversed progression of plaques by 5.9% in a small subset of patients with high plaque burden. Before the ASTEROID trial, most cardiologists thought that stabilization in plaque volumes was the best that could be achieved. Crestor changed that perception. Now, some physicians believe they can turn back the clock on heart disease, based on Crestor results.

The ASTEROID trial also established that Crestor lowered levels of LDL by more than 53%, and it raised levels of HDL by nearly 15%. These changes in cholesterol levels seen in the latest study were the largest ever found in a major monotherapy trial of a statin. Although the trial did not answer the question of whether decreased plaque volume would translate to fewer heart attacks and strokes, Crestor did reverse atherosclerosis, a condition that develops over many years. In addition, the study did not address whether that made a difference in reducing the risk of a heart attack or stroke. It is generally believed that statins are thought to exert their powerful effects in reducing heart attack risks by making plaque more stable, not necessarily shrinking its size.[20]

Regardless of these caveats, the ASTEROID trial distinguished Crestor from other statins and reaffirmed a growing body of evidence supporting the mantra for LDL-cholesterol levels: "lower is better." Nowadays, some cardiologists even target LDL-cholesterol levels lower than 70–80 mg/dL for high-risk patients.

Cholesterol Drugs beyond Statins

Since Mevacor became available on the market in 1987, the statins have revolutionized the treatment of hypercholesterolemia. By 2005, it was estimated that more than 56 million people have benefited from some form of cholesterol reduction. Because statins are not suitable for all patients, many other mechanisms have been explored to modulate the level of cholesterol. They include fibrates, cholesterol ester transfer protein (CETP) inhibitors, peroxisome proliferator–activated receptor (PPAR) agonists, nicotinamide receptor agonists as exemplified by nicotinic acid, and acyl-CoA: cholesterol acyltransferase (ACAT) inhibitors.

One biotech company, Isis Pharmaceuticals, recently reported a successful midstage trial for a cholesterol-lowering drug, but the technology to make that drug, known as antisense technology, has had a troubled history. Microbia (whose name was changed to Ironwood Pharmaceuticals in April 2008), a privately held biotechnology company in Cambridge, Massachusetts, has a drug in early clinical trials that blocks the absorption of cholesterol from food. That is the same approach used by Zetia (ezetimibe), a drug sold by Schering-Plough. Vytorin, the combination of Zetia and Zocor jointly sold by Schering Plough/Merck, is also now available.

Zetia and Vytorin

Modern drugs are increasingly being discovered through "rational drug design." In contrast, Zetia, a novel cholesterol absorption inhibitor, was discovered through a classical in vivo animal model activity optimization without a clear understanding of the molecular target.

Schering-Plough introduced ezetimibe (Zetia) to the market in 2002. Unlike the statins, which inhibit HMG-CoA, an enzyme needed in the liver to produce cholesterol, Zetia blocks absorption of cholesterol in food from the small intestine into the bloodstream. It was the first of a new class of LDL-cholesterol–lowering agents, so-called intestinal cholesterol absorption inhibitors. In addition to its mechanistic novelty, Zetia has a long duration of action, low systemic exposure, and an excellent safety profile. The discovery path to ezetimibe encompassed the common themes of scientific achievement: inspiration, perspiration, and serendipity.[21] However, Zetia is not as effective as statins. It reduces LDL-cholesterol levels by just 18–20%. People can achieve similar reductions on their own through diet and exercise. So the drug is being positioned mainly as a complement to statins to boost their effectiveness.[22]

The discovery of Zetia is another story of serendipity in the annals of drug history. In 1990, Schering-Plough began to work on ACAT inhibitors to treat atherosclerosis. Duane A. Burnett and coworkers worked on 2-azetidinones—a class of compounds containing rigid four-membered rings.[23] In the early stages of the project, scientists noticed a disparity: although the compounds did not possess in vitro ACAT inhibitory potency, they had significant cholesterol absorption reduction in an in vivo animal model using cholesterol-fed hamsters. Puzzlingly, when they gave an early prototype to humans, they found that the 250-mg daily dose was not much more efficacious than the 25-mg starting dose.

Because the ACAT assay was not indicative of actual LDL-cholesterol reduction, the medicinal chemists opted to use the onerous route of establishing structure-activity relationships (SAR) guided solely by the *in vivo* animal model. The absence of an *in vitro* assay clearly made the endeavor extraordinarily challenging for the medicinal chemists.[24] To explore the complex SAR, metabolism-blocking compounds were prepared. Ultimately, ezetimibe was discovered as the fruit of exemplary coalescence of synthetic organic chemistry, medicinal chemistry, pharmacology, and drug metabolism. It was shown to be a highly potent inhibitor of cholesterol absorption in several animal models, including hamsters, dogs, and monkeys. In a clinical trial with patients who were pre-equilibrated with a low-cholesterol diet, treatment with 10 mg of ezetimibe reduced LDL-cholesterol levels by an additional 18.5%. As an added benefit, it also exhibits desirable effects on levels of triglycerides, apolipoprotein B, and C-reactive protein. A joint venture between Merck and Schering-Plough

Pharmaceuticals was responsible for the development and marketing of ezetimibe. The FDA approved ezetimibe in October 2002 after a 10-month review, and Merck/Schering-Plough Pharmaceuticals began selling it with the trade name Zetia.[25]

In 2004, the American Chemical Society recognized chemistry's heroes. Among them was the team of medicinal chemists responsible for the birth of Zetia, a drug that represents the first new approach to reducing cholesterol levels since the discovery of statins: Duane A. Burnett, John W. Clader, Brian A. McKittrick, Stuart B. Rosenblum, and Sundeep Dugar.

Because no significant effects on liver enzyme levels have been observed in animals or during the clinical trials of ezetimibe, it is ideal for combination therapies with a statin. Therefore, Merck/Schering-Plough subsequently developed a fixed combination of Zetia and Zocor, which was approved by the FDA in July 2004. The combined drug is sold under the trade name Vytorin,[26] which quickly became their best-selling product.

Merck and Schering-Plough have made Vytorin the fastest-growing brand name in the $22 billion-a-year cholesterol reduction market. They promote the fact that Vytorin reduces cholesterol in two ways: metabolism and diet. They have distinguished Vytorin from other cholesterol-lowering drugs by emphasizing studies showing that it achieved a greater reduction of LDL-cholesterol levels than Lipitor, Zocor, or Crestor. Vytorin has also extended the effective proprietary life of Zocor and has successfully competed in the crowded class.

Cholesterol pills, benefiting from television advertising, helped drug makers recover from the end of patent protection on top-selling medicines such as Schering-Plough's Claritin (loratadine; patent expired in 2002) and Merck's Zocor (patent expired in June 2006).[27] The sales of Zetia and Vytorin in 2006 were $3 billion, saving Schering-Plough from the red and helping Merck pick up the slack caused by the Vioxx lawsuit and the Zocor patent expiration.

On January 14, 2008, in a cruel twist of fate, Merck & Co. and Schering-Plough Corp. announced that a long-awaited trial showed their cholesterol drug Vytorin failed to slow progression of heart disease better than a cheaper drug, the generic simvastatin (Zocor), threatening the companies' $5 billion-a-year cholesterol-fighting franchise.[28] The trial, named the Effect of Combination Ezetimibe and High-Dose Simvastatin vs. Simvastatin Alone on the Atherosclerotic Process in Patients with Heterozygous Familial Hypercholesterolemia (ENHANCE) trial, involved 720 patients and lasted two years. Schering-Plough Chief Executive Fred Hassan said ENHANCE was "not a large trial" and was "in a very, very special" population. He added, "I don't know why this would have any impact on mainstream use" of Vytorin. On the other hand, Dr. Steven Nissen, the chairman of cardiology at Cleveland Clinic, recommended "physicians should now stop using (Zetia) or Vytorin as a primary therapy for patients with high cholesterol."[28] This was the first time that lowering cholesterol levels did not translate into less artery clogging. This turn of events challenges the conventional wisdom, indicating that lower is not always better. Maybe the statins work through a different mechanism, such as anti-inflammation, to prevent heart diseases? We have yet to find out what will transpire from the ENHANCE trial regarding the role of Zetia in the future.

In fairness, both opinions are probably extremes. A happy medium is likely a fairer assessment. Some cardiologists said they would stick with Zetia, at least for certain patients—for instance, those who are at risk of side effects from a high dose of statin.

"We treat based [on] goals rather than [on] specific drugs," said Michael Davidson, director of preventive cardiology at the University of Chicago's Pritzker School of Medicine. "That's the important message that is not changed by this trial." Dr. Davidson worked on some of the trials whose data supported FDA approval of Vytorin and Zetia.[28]

In early April 2008, a panel of cardiologists at the annual meeting of the American College of Cardiology dealt a blow to Zetia and Vytorin. The panel called on physicians to sharply curtail their use of those two cholesterol drugs. The news sent Schering-Plough's stocks to a tailspin, dropping nearly 29% to $14.41 a share because sales of both Zetia and Vytorin accounted for 60% of Schering-Plough's profit. In response, the company announced plans to cut $1.5 billion by 2012 by slashing its workforce by 10%, translating to 5,500 employees. On the other hand, the blow to Merck was relatively mild, with stocks down 15% to $37.95 a share. And Zetia and Vytorin were responsible for only 15% of Merck's profit.

Torcetrapib, a bitter disappointment

Statins like Lipitor and Zocor have been on the market for 20 years. They have been used by millions of Americans and have reduced heart attacks by about one-third. Statins work by reducing the levels of LDL cholesterol that form plaque on the walls of arteries. In addition, they also slightly increase the levels of good cholesterol, HDL cholesterol. Unfortunately, statins boost the levels of HDL cholesterol by only about 5%. While LDL cholesterol is a risk factor for coronary heart disease, its cousin, HDL cholesterol, is a protective factor. Increasing the good cholesterol could potentially offer an entirely new way to help prevent heart attacks. Good cholesterol makes it more difficult for bad cholesterol to form plaques, which is why a patient's cholesterol risk is better expressed as a ratio of total cholesterol to good cholesterol. The smaller the ratio, the better.

As early as 1977, the famous Framingham Heart Study had established that high levels of HDL cholesterol actually protect against heart disease in many participants.[29] Even though this was not conclusive evidence that raising HDL cholesterol would reduce a patient's risk for heart disease, many other studies also suggested that high levels correspond to a lower incidence of coronary heart disease. In the United States, average HDL cholesterol is about 45 mg/dL in men and 55 mg/dL in women. An HDL-cholesterol level less than 40 is an especially bad sign, while anything greater than 60 is considered good. Somehow, estrogen levels correlate to the HDL-cholesterol levels—a fact that might explain why women have relative fewer incidences of coronary heart disease and generally live longer than men.

So far, high-dose niacin (nicotinic acid) is the only drug available to raise HDL-cholesterol levels. But its impact is modest, increasing the good cholesterol by only about 14% for men and 20% for women, and its side effects, such as itching and hot flashes, are bothersome. Kos, a company that sold nicotinic acid with the trade name Niaspan, was acquired by Abbott Laboratories at the end of 2006. In July 2006, AstraZeneca and Abbott announced that they would jointly develop and commercialize a fixed-dose combination product that would target three blood lipids: LDL, HDL, and triglycerides. This single-pill product would contain AstraZeneca's Crestor and Abbott's TriCor, a fibrate drug. In addition, Merck tested Cordaptive, which combines nicotinic acid with laropiprant, a molecule to reduce the flushing side effects. Regrettably, the FDA rejected Cordaptive in April 2008.

Another tactic for boosting the HDL-cholesterol levels is to block the cholesterol-ester transfer protein (CETP), which is an enzyme in the blood that shuttles cholesterol and triglycerides between LDL and HDL. The concept owes its genesis to a group of Japanese scientists.

In the late 1980s, physician Yuji Matsuzawa and his colleagues at Osaka University Medical School reported a unique epidemiologic observation.[30] CETP deficiency, caused by mutation of the CETP gene, was extremely frequent in Omagari City in Japan. There, the mutation was more than 20 times more prevalent, and the number of individuals with plasma HDL cholesterol greater than 100 mg/dL was 5–10 times higher than in other areas of Japan. This discovery led to the hypothesis that blocking the action of CETP would increase levels of HDL cholesterol.

Scientists at Pfizer read Matsuzawa's article in the *New England Journal of Medicine*[31] and speculated that CETP inhibition could be a therapeutic strategy for raising HDL-cholesterol levels, although the concept was controversial at the time. This was precisely the kind of product the general public wanted the industry to focus on: not another "me too" drug that achieves a merely incremental advance over some existing therapy, but a completely new approach that could spur a huge leap forward in the battle against heart disease. Pfizer commenced a high-throughput screening effort but found only a few compounds that were weak CETP inhibitors.[32] Although initial efforts were disappointing, a breakthrough came in 1993, when a young chemist, Roger B. Ruggeri, just hired by Pfizer, was assigned to check a final list of 10 possibilities. One of the two chemicals he chose panned out, giving rise to torcetrapib, a small-molecule CETP inhibitor that sharply increased levels of HDL in rabbits.[33] More important, the blockage of CETP was associated with vastly lower levels of arterial plaques in those rabbits. The rabbit studies were the tiebreaker, which justified moving the drug torcetrapib forward with human tests.

In one human trial, researchers at the University of Pennsylvania and Tufts University gave 19 patients a 120-mg torcetrapib pill daily for four weeks. Ten of the 19 patients also took Lipitor. They found that the drug doubled HDL cholesterol in people with worrisomely low levels of the HDL cholesterol. It also reduced LDL cholesterol. HDL levels were elevated an average of 46% in those taking just torcetrapib and jumped 61% in those taking both torcetrapib and Lipitor. Additionally, six patients took the torcetrapib pill twice a day in the study's third phase, and their HDL jumped 106%.[34]

In a phase II study involving 500 patients, those who received torcetrapib (60 mg) and Lipitor (10, 20, 40, and 80 mg) had increases in

HDL-cholesterol levels of 44–66%. At the same time, their LDL cholesterol dropped 41–60%. Some cardiologists had raised concerns about torcetrapib, noting that the drug raised blood pressure in many patients, a serious side effect for a heart medicine. Patients taking 60 mg of torcetrapib with Lipitor experienced an increase in systolic blood pressure of approximately 3 mmHg. Because the patient population was relatively small, it would take a much larger sample size to determine if the hypertension adverse effect was general for torcetrapib.

In 2004, Pfizer commenced the most expensive phase III clinical trial in history for torcetrapib, involving 15,000 patients with a cost greater than $800 million.[35] In the so-called ILLUMINATE trial, 7,500 patents were given a combination of torcetrapib and Lipitor, and another 7,500 patents were given Lipitor alone. Pfizer had hoped that the trial would demonstrate that people taking the combination pill would be significantly less likely to suffer deaths or heart problems than those taking Lipitor alone. In December 2006, an independent monitoring panel found that patients receiving torcetrapib were dying at a higher rate and had more heart problems than patients who did not receive the drug. Eighty-two people had died in the clinical trial, versus 51 people in the same trial who had not taken it. Not only were there 31 more deaths among the people taking torcetrapib, but similar discrepancies were noted in the number of patients suffering heart failure and other problems. The panel recommended terminating the trial, and Pfizer promptly told more than 100 trial investigators to stop giving patients the drug.[36]

For the estimated tens of millions of Americans at risk for heart disease, Pfizer's halting of its trials of torcetrapib was doubly disappointing. Meanwhile, in a trading frenzy, Pfizer's stocks dropped nearly 11%, shaving more than $21 billion off its market value.

Two weeks after Pfizer's termination of the trials for torcetrapib, Merck halted its phase III trials for CETP inhibitor anacetrapib, which had completed phase IIB study. But Merck resumed its phase III trials of anacetrapib in 2008, assuming that torcetrapib's toxicity is molecule-based rather than mechanism-based. Roche Holdings, the Swiss pharmaceutical company, was developing the CETP inhibitor JTT-705 with Japan Tobacco and Bayer. They analyzed torcetrapib data before advancing their drug that was scheduled to enter phase III study.[37] Interestingly, in early April 2008, Roche said that its CETP inhibitor R1658 did not have adverse effects on patients' blood pressure.

At this point, Pfizer and independent cardiologists must now determine whether torcetrapib's failure means that all medicines to raise good cholesterol will have similar problems (mechanism-based toxicity), or the problem was specifically related to some defect in torcetrapib (molecule-based toxicity).[38] It is also a sobering reminder that pioneering drug research is a risky business, both for patients who take unproven drugs in clinical trials and for companies that spend a substantial portion of their research budgets on them.

Similar to the ENHANCE trial, the spectacular failure of torcetrapib challenges the conventional wisdom that raising the levels of HDL per se lowers the risk of heart attacks. Meanwhile, many now begin to believe that LDL cholesterol is a marginal risk factor. It is not unlikely that statins work by reducing inflammation, which is now considered a risk factor of heart disease.

ApoA-1 Milano and Roger Newton

Years ago when the field of drug discovery was still in its infancy, it was not uncommon for a single scientist to be responsible for the birth of numerous drugs. Leo Sternbach at Hoffmann-La Roche invented Librium (chlordiazepoxide), Valium (diazepam), and half a dozen other benzodiazepine tranquilizers. Paul A. J. Janssen was associated with the discovery of haloperidol, fentanyl, risperidone, and approximately 80 other drugs! Unfortunately, those days are long gone, and most scientists in the drug industry would be very fortunate to be associated with even one drug nowadays. In an uncommon feat, Roger Newton, after playing a key role in the Lipitor saga, moved on to develop another significant heart drug.

Newton is quite outspoken, a trait that served him well as the most vocal champion of Lipitor. Cathy Sekerke, a biologist on the Lipitor team, said: "Without Roger, Lipitor would have been killed many times over earlier."[39] However, his philosophy of "let the science speak, let the data decide" did not always serve him well politically. When Parke-Davis licensed several cardiovascular drugs from other companies, Newton was ferociously against them because he did not think they were good drugs. Despite his tremendous contributions to the birth of Lipitor, his honest criticism did not win him favors with management. He was treated badly by his bosses. In 1995, he was a senior director of cardiovascular biology with more than 60 biologists under him. In 1996, he was stripped of all managerial responsibilities and given a title of "Distinguished Research Fellow," with three postdoctoral fellows reporting to him. The message to him was loud and clear: he fell out of grace with the managers, no longer favored from above. He once obtained a director-level position at Bristol-Myers Squibb, but decided to stay at Parke-Davis for two additional years just to shepherd Lipitor to the market.

In the early 1990s, Newton began advocating the approach of increasing HDL, but each time his idea was dismissed. Shortly after Lipitor's launch at the beginning of 1997, Newton approached management for the third time to propose working on his HDL idea. He was again rejected. He told himself: "Three strikes, I'm out" and then began to execute his exit strategy.[40] In May 1998, Newton and three fellow colleagues, Tom Rea, Michael Pape, and Charlie Bisgaier, left Parke-Davis to start a new company called Esperion Therapeutics Inc., whose mission was to bring a novel approach to the emerging area of HDL therapy and reverse lipid transport for the acute treatment of cardiovascular disease. Newton became the chief executive officer. The chairman of Esperion's Board, David Scheer, helped with the business end, which allowed Newton to focus on the science. Esperion started with $15.5 million and raised about $42 million in venture capital; the remainder of the financing was equity in the amount of $150 million.

One of the drugs that Esperion developed, ETC-216, would catch the attention of the world in general, and of Pfizer in particular. The story of ETC-216 goes back to Italy in 1974.

University of Milan researchers in 1974 found that a man named Valerio Dagnoli, from the lakeside town of Limone sul Garda in northern Italy, had a very low level of HDL, which keeps arteries from accumulating brittle plaque that can break off and choke blood flow. Dagnoli also had high levels of triglycerides, a bad form of fat in blood that can lead to cholesterol-plaque deposits in the arteries. Despite this highly abnormal lipid profile, the middle-age man had no evidence of cardiovascular disease, and his parents had also enjoyed long healthy lives.[41]

When blood tests were done on the entire 1,000 inhabitants of Limone, about 40 individuals were found to have a similar lipid abnormality. Using birth records maintained in the local church going back several hundred years, it was found that these 40 individuals were all traceable to common ancestors from 1780 (Giovanni Pomaroli and Rosa Giovaneli). This then led to the discovery that these 40 individuals had a genetic mutation in the gene that makes a protein called ApoA-1, which becomes a part of the HDL-cholesterol particle. In the 1990s, Dr. P. K. Shah of Cedars-Sinai Medical Center in Los Angeles injected a synthetic version of this variant HDL into rabbits and mice. He showed that ApoA-1 Milano not only reversed plaque buildup but also stabilized and reduced inflammation of what was left.

In 1996, the Swedish drug firm Pharmacia and Kalamazoo-based Upjohn collaborated with the University of Milan to test ApoA-1 Milano for the treatment of atherosclerosis. Two years later, Pharmacia decided to out-license the drug since its focus was small molecules and ApoA-1 Milano is a biologic. Newton and his team licensed ApoA-1 Milano from Pharmacia soon after founding Esperion.[42] That compound became known as ETC-216, where ETC stands for Esperion Therapeutics compound.

Running a small biotechnology firm is not without its perils. In March 2002, Esperion stock was at an all-time low. Newton had to lay off a quarter of his workforce, one of the hardest things he had to do. But things began to look up when positive phase II results were published. Newton then persuaded Dr. Steven Nissen of the Cleveland Clinic to conduct studies on ETC-216 using intravascular ultrasound—a bioimaging method that uses a tiny ultrasound probe inserted into the coronary artery to directly image the size of atherosclerotic plaques. These phase II results showed a statistically significant reduction in plaque volume in patients with acute coronary syndrome at the end of six weeks of infusions. It was revolutionary! Statin drugs, which lower LDL cholesterol that causes plaque in the first place, reduce the risk of dying from heart disease only 30% or so. By targeting HDL as well, the risk might be halved. ETC-216 needs to be administered by a physician through intravenous infusions. Since ETC-216 is an HDL mimetic (meaning it acts like HDL, only better), it will actually allow us to bring about a regress of vulnerable plaques, rapidly eliminate excess cholesterol, and thereby reduce the risk of heart attacks. Therefore, it complements other therapies such as statins.

With the positive results from phase II trials, ETC-216 was poised to be licensed to a big pharmaceutical company for additional phase II and III development since Esperion did not have the financial means or the resources to carry out such large-scale trials. According to the contract signed between Pharmacia and Esperion, Pharmacia had the first right to decide if it wanted to license. Pharmacia declined to buy the right. As a consequence, other companies would be able to acquire Esperion without legal ramifications. In fact, Merck and some other big pharmaceutical companies initiated conversations with Esperion. In order to prevent others from getting their hands on ETC-216, Pfizer acquired Esperion for $1.3 billion in December 2003. This price represented a 54% premium over Esperion's average closing share price during its last 20 trading days. It was the 15th largest transaction in the pharmaceutical industry that year. It will take some years to learn the fate of ETC-216, after its phase II and III results become available. In January 2007, Pfizer announced that it would close its Ann Arbor site, along with Newton's Esperion. In May 2008, Newton raised $22.75 million to buy back Esperion Therapeutics from Pfizer.

The field of cholesterol treatments is ever expanding. In addition to statins, niacin, fibrates, resins, cholesterol absorption inhibitors, CETP inhibitors, and ApoA-1 Milano, scientists around the globe are also working on many other mechanisms for reducing LDL cholesterol and/or elevating HDL-cholesterol levels. For instance, Isis Pharmaceuticals has an antisense drug, ISIS-301012, that muffles the gene for a protein called apoliprotein B-100, or ApoB. So far, statins are superior in terms of both efficacy and risk–benefit profile. However, statins only manage one of the major risk factors, LDL cholesterol. To eradicate heart disease, a multi-pronged attack needs to be launched. For instance, smoking represents the single most import risk factor for heart disease. The antismoking campaign was responsible for the plummeting rate of heart disease in the 1970s. In addition, sedentary lifestyle, hypertension, and diabetes are also important risk factors in the development of coronary heart disease. Removing these risk factors is just as important as keeping LDL-cholesterol levels at bay.

CHAPTER 8

Reflections

Triumph of the Heart

The story of statins is a success story for science (both basic and applied) and scientists (in both academia and industry). It contains one of the classic scientific and marketing battles in the history of the pharmaceutical industry. More important, it has been a great boon for the millions of patients who have benefited from statins in preventing coronary heart disease. The story of the statins is a triumph of the heart.

Statins, a class of cholesterol-lowering drugs, have revolutionized the landscape of coronary heart disease treatment. Since Merck's marketing of Mevacor in 1987, the world has benefited from statins in numerous ways.[1] As a class of drugs, statins have set standards on numerous fronts in helping manage LDL cholesterol, one of the major risk factors for coronary heart disease. Statins set a high standard in efficacy, a high standard in safety, and a high standard in financial success for the patients, payers, and the pharmaceutical industry.

Not only do statins greatly reduce cholesterol and lower mortality in people at risk for heart attacks, but some studies also suggest that they might help prevent or treat a wide range of ailments, including Alzheimer's disease,

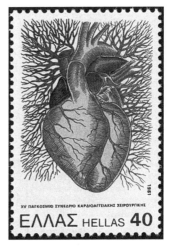

Figure 8.1 The Heart © Greek Post.

multiple sclerosis, bone fractures, some types of cancer, macular degeneration, and glaucoma.[2,3] The world has already benefited from the statins in many ways.

Statins set a high standard in efficacy

Low is good, but lower is even better. Fifty years ago, the connection between cholesterol and coronary heart disease was still in question. Twenty years ago, the merit of lowering LDL cholesterol was not even unanimously agreed upon. Cholesterol drugs before the statins, such as resins, niacin, and fibrates, worked to some extent but were also seriously limited by their side effects. Thanks to the emergence of the statins, with Mevacor as the first on the market in 1987, all these questions on the relationship between cholesterol and coronary heart disease are answered beyond any shadow of doubt. Today, the statins have annual sales of more than $20 billion. Hundreds of millions of patients have benefited from statins by delaying and even preventing coronary heart disease.

Most statins do not actually remove preexisting plaques; rather, they stabilize them and stop plaque progression (although Crestor has shown some promise of reversing the formation of plaques). Therefore, arteries are less likely to rupture. Invariably, trial after trial has shown beneficial effects after three to six months of therapy.

Thanks largely to the statins, the average total cholesterol level for Americans has dropped from 222 mg/dL in 1960 to 203 mg/dL in 2003. The past decade saw levels of LDL cholesterol decrease by 4.5%, from 129 to 123 mg/dL. The level of cholesterol drop could translate into prevention of thousands of heart attacks and strokes and thousands of lives saved. From the Framingham Study and other epidemiologic studies, we now know that a 1% reduction of LDL cholesterol translates into a 1–2% reduction in the risk of a heart attack or other serious cardiovascular event. Indeed, annual deaths from heart disease in the United States dropped from 800,000 in the 1980s to 650,000 in 2002. Not only have annual hospital admissions for heart attacks fallen more than 6% since 2000, but admissions for other conditions related to coronary artery disease are leveling off or shrinking, as well.[2] Because of statins, the savings from hospitalization and surgeries such as bypass and angioplasty are enormous.

In addition to lowering LDL cholesterol, the statins have six additional benefits. They have been shown to lower C-reactive protein (CRP), a sign of inflammation; reduce risk of stroke; treat autoimmune disease and prevent transplant rejection; improve bone health; fight dementia and Alzheimer's disease; and reduce the risk of diabetes. No panacea exists today, but statins are getting really close. No wonder cardiologists often joke that statins should be added to the water supply.

Statins set a high standard in safety

Overall, statins have proved to be remarkably safe. They are proving ever more useful and have saved many lives, particularly among middle-age men at risk for heart disease, the group most widely studied. This is, of course, not to say that statins are absolutely safe. No medications, not even over-the-counter drugs like aspirin and acetaminophen, are 100% safe.

The most conspicuous adverse effect of the statins is rhabdomyolysis, which seems to manifest in different degrees for the different statins. While most statins have extraordinarily low chances of producing rhabdomyolysis when taken in approved doses, Baycol appears to have a much higher frequency. Most statins, including Lipitor, Zocor, Pravachol, and Lescol, are all contraindicated with fibrates such as Lopid and Tricor.

There have been scattered reports of adverse effects such as minor muscle pain, liver failure, and weakened cognitive function when taking statins. However, many studies have demonstrated that these adverse effects take place just as frequently when the patients take a placebo.[4] Undoubtedly, statins have a preferable risk–benefit profile, which sets a high standard for safety for future heart drugs in general and cholesterol drugs in particular.

Statins set a high standard in financial benefits

Especially in recent times, the cost of drugs has been a contentious issue. Despite the increasing costs of pharmaceuticals, patients taking pills actually save money because they avoid surgeries and hospitalizations. Each year, the preventive effects of statins allow patients to avoid 64,600 admissions to the emergency department, which translates into tremendous cost savings for U.S. citizens. In the end, statins save money for the patients and payers.

The pharmaceutical industry is among the very few industries that contribute in a positive way to the balance of trade for the American economy, its financial success helping to offset the fact that America imports much more than it exports. Statins represent the golden age of America's pharmaceutical industry. The statins have done so much for patients, pharmaceutical company employees and shareholders, and even the American economy that it will be difficult to find the next class of drugs to fill the revenue void after statins' patent expirations.

By 2007, several statins had lost their patents, allowing generic versions to be sold, thus becoming much less expensive than the brand names. For patients with fewer risk factors for coronary heart disease, generic lovastatin (Mevacor), fluvastatin (Lescol), pravastatin (Pravachol), and simvastatin (Zocor) might suffice. Patients with multiple risk factors can still use Lipitor and Crestor to achieve even more dramatic reduction of LDL cholesterol. With either generic or brand-name statins, both patients and payers still save money compared with no statin therapy, because the surgeries and hospitalizations are significantly more expensive than the medicines.

The Drug Industry and the Blockbuster Model

The 1980s and 1990s were an era of blockbuster drugs that represented the golden age of the drug industry. Largely thanks to statins, each a blockbuster drug, the pharmaceutical industry enjoyed unprecedented prosperity during the first decade of the twenty-first century. The discovery, production, and marketing of Lipitor, a $13 billion drug, was a combination of serendipity, determination, timing, and marketing. So many factors had to come together to make Lipitor a success that any encores are likely to be few and far between. The drug industry has become a victim of its own success—in order to match the profits generated from so many blockbusters, there must be a profusion of new blockbusters to market, but this has not yet come to pass. Among many reasons for the blockbuster drought are the difficulty of improving existing drugs, tighter approval standards, and big company red tape. Indeed, blockbuster drugs target common ailments such as cholesterol, hypertension, pain, and depression. Most of these already have rather good drug therapies.

In light of all available evidence, the blockbuster model of drug discovery does not seem to work anymore, simply because if you operate by the blockbuster model, any setback is going to be catastrophic. The final unexpected failure of torcetrapib was a crushing blow, exemplifying the peril of wagering too much of a company's future on a single product, as opposed to spreading the risk among several smaller drugs. The current era of drug discovery is changing in ways that favor small, nimble biologics using biotechnology.[5]

The year 2006 was an extremely challenging one for the pharmaceutical industry. The American drug industry, the envy of the world, is still going through a tough time. Patent expirations afflicted most, if not all, major drug firms. Drugs that collectively are worth between $23 billion (in the U.S.) and $28 billion (worldwide) in annual sales lost exclusivity in 2006. Merck lost its patent protection for Zocor, a drug with $5.2 billion in sales, in 2004, and that same year Bristol-Myers Squibb also lost its patent protection for Pravachol. As a consequence, patients can buy cheaper generics, and the brand-name companies lost a revenue source worth billions of dollars. In response, the drug industry handed out 25,000 pink slips through early retirements and layoffs in 2006 alone.

Today, the future of the drug industry certainly looks grim, as so many blockbusters, many of them statins, are losing their patent protection. But on the positive side, historically scientists in the pharmaceutical industry have a knack of creating waves of newer and better drugs as science progresses. Moreover, demographics are the best thing the industry has going forward. People are living longer, largely thanks to the past success of drug discovery. Demands for pharmaceuticals should be on the rise over the next 15 years as the baby boomers reach their senior years. Past successes have prolonged people's lives and perpetuated the notion that that drugs can be made to address every medical need. Thus, the industry has, in a way, created future needs to be met, by meeting the needs of the present.

APPENDIX

Drug Names

Trade Name	Generic Name	Company
Accupril	Quinapril	Parke-Davis
Altace	Ramipril	Hoechst
Atromid-S	Clofibrate	Imperial Chemical Industries
Aureomycin	Chlortetracycline	Lederle
Baycol	Cerivastatin	Bayer
Benadryl	Diphenhydramine	Parke-Davis
Caduet	Atorvastatin calcium/ amlodipine besylate	Pfizer
Camptosar	Irinotecan	Pfizer
Capoten	Captopril	Bristol-Myers Squibb
Celebrex	Celecoxib	Pfizer
Chloromycetin	Chloramphenicol	Parke-Davis
Cipro	Ciprofloxacin	Bayer
Claritin	Loratadine	Schering-Plough
Colestid	Cholestipol	Upjohn
Crestor	Rosuvastatin	AstraZeneca
Dalzic	Practolol	Zeneca
Diflucan	Fluconazole	Pfizer
Diuril	Chlorothiazide sodium	Merck
Dolobid	Diflunisal	Merck
Fozitec	Fosinopril sodium	Squibb
Geodon	Ziprasidone	Pfizer

Trade Name	Generic Name	Company
Inderal	Propranolol	Wyeth-Ayerst
Inhibace	Cilazapril	Roche
Januvia	Sitagliptin phosphate	Merck
Lescol	Fluvastatin sodium	Novartis
Lipitor	Atorvastatin calcium	Pfizer
Lopid	Gemfibrozil	Parke-Davis
Lotensin	Benazepril	Ciba-Geigy
Lotrimin	Miconazole	Janssen
MER/29	Triparanol	Richardson-Merrell
Mevacor	Lovastatin	Merck
Niaspan	Nicotinic acid	Kos
Nizoral	Ketoconazole	Janssen
Norvasc	Amlodipine besylate	Pfizer
Pravachol	Pravastatin sodium	Bristol-Myers Squibb
Prozac	Fluoxetine hydrochloride	Lilly
Questran	Cholestyramine	Bristol-Myers Squibb
Tagamet	Cimetidine	GlaxoSmithKline
Terramycin	Oxytetracycline	Pfizer
Terramycin	Tetracycline	Pfizer
Tofranil	Imipramine	Pfizer
Tricor	Fenofibrate	Abbott
Trosyl	Tioconazole	Pfizer
Vasotec	Enalapril	Merck
Vfend	Voriconazole	Pfizer
Viagra	Sildenafil citrate	Pfizer
Vytorin	Ezetimibe/simvastatin	Schering-Plough/ Merck
Zantac	Ranitidine	GlaxoSmithKline
Zestril	Lisinopril	Merck
Zetia	Ezetimibe	Schering-Plough
Zithromax	Azithromycin	Pfizer
Zocor	Simvastatin	Merck
Zoloft	Sertraline hydrochloride	Pfizer
Zyrtec	Cetirizine dihydrochloride	Pfizer

NOTES AND REFERENCES

PREFACE

1. Bronowski, Jacob, in the Preface of *The Ascent of Man*, Little Brown: New York, NY. (1973).

PROLOGUE

1. Berenson, Alex, "Patent challenge to Pfizer is rejected," *New York Times*, December 17, 2005.
2. Under Hank McKinnell's reign, Pfizer lost roughly $137 billion in market value. Yet when he was forced by the board to retire in December 2006, McKinnell walked away with a $200 million retirement package. That incited a *New York Times* editorial with the title "A Lump of Coal Might Suffice"—the perfect Christmas gift from the shareholders who lost money and from laid-off employees (Gretchen Morgenson, *New York Times*, December 24, 2006).

CHAPTER 1

1. Reinitzer, Friedrich, "Contributions to the knowledge of cholesterol," *Monatshefte für Chemie* 1888, 9, 421–441. Its English translation appeared in *Liquid Crystals* 1989, 5(1), 7–18.
2. Reinitzer's 1888 landmark paper was groundbreaking not only for his correct assignment of cholesterol's formula, but also because he documented the existence of a liquid crystal for the first time in history. After he treated cholesterol with benzoic acid anhydride, Reinitzer obtained cholesteryl benzoate, whose properties made the reserved scientist gush in his description: "*The most beautiful crystals* are obtained by slow evaporation of a solution in ether with as much alcohol as it can tolerate without clouding" (Reinitzer, "Contributions"). Reinitzer found that when he melted cholesteryl benzoate, it became a cloudy liquid and then cleared up as its temperature rose. After cooling, the liquid turned blue before crystallizing, a characteristic of a liquid crystal. Reinitzer is now considered the grandfather of liquid crystal science. While we watch our wonderful liquid crystal display (LCD) televisions, we all owe Reinitzer a debt of gratitude.

3. Laylin, James K. (ed.), *Nobel Laureates in Chemistry, 1901–1992*, American Chemical Society: Washington, DC, 1993, pp. 164–173.
4. Vance, Dennis E.; Bosch, Henk Van den, "Cholesterol in the year 2000," *Biochimica et Biophysica Acta* 2000, 1529, 1–8.
5. Williams, T., *Robert Robinson, Chemist Extraordinary*, Clarendon: Oxford, 1990.
6. Barton, Derek H. R., *Some Recollections of Gap Jumping*, American Chemical Society: Washington, DC, 1991, p. 28.
7. Morris, Peter J. T.; Bowden, Mary Ellen, "Robert Burns Woodward: A biographical introduction," in Benfey, O. T.; Morris, P. J. T. (eds.), *Robert Burns Woodward, Architect and Artist in the World of Molecules*, Chemical Heritage Foundation: Philadelphia, 2001, pp. 3–10.
8. Mulheirn, Greg, "Robinson, Woodward and the synthesis of cholesterol," *Endeavor* 2000, 24(3), 107–111.
9. Sadly, Woodward died prematurely at the age of 62. Frank H. Westheimer, a chemistry professor at Harvard since 1953 and a friend of Woodward, wrote: "In addition to his great contribution to rational synthesis he and Roald Hoffmann, after a conversation with E. J. Corey, formulated the principle of the conservation of orbital symmetry, a set of principles that, for example, control the formation and cleavage of rings. This discovery also had an enormous impact on the synthesis, supplying an additional important intellectual component to the discipline" ("Robert Burns Woodward, scientist, colleague, friend," in Benfey and Morris, *Robert Burns Woodward*, p. 16).
10. Friedman, Meyer; Friedman, Gerald W., *Medicine's 10 Greatest Discoveries*, Yale University Press: New Haven, 1998, pp. 153–167; Raab, W., "Nikolai Nikolaievitch Anitchkov (1885–1964)," *Cardiologia* 1965, 47(3), 207–208.
11. Klimov, A. N.; Nagornev, V., "Evolution of cholesterol concept of atherogenesis from Anitchkov to our days," *Pediatric Pathology and Molecular Medicine* 2002, 21(3), 307–320.
12. Steinberg, Daniel, "Thematic review series: The pathogenesis of atherosclerosis. An interpretive history of the cholesterol controversy: Part I," *Journal of Lipid Research* 2004, 45(9), 1583–1593.
13. "The plowboy interview: John Gofman, nuclear and antinuclear scientist," *Mother Earth News*, March/April 1981. Available at: http://www.ratical.org/radiation/CNR/PlowboyIntrv.html.
14. Steinberg, Daniel, "Thematic review series: The pathogenesis of atherosclerosis. An interpretive history of the cholesterol controversy: Part II: The early evidence linking hypercholesterolemia to coronary disease in humans," *Journal of Lipid Research* 2005, 46(2), 179–190.
15. Kennedy, Eugene P., "Hitler's gift and the era of biosynthesis," *Journal of Biochemistry* 2001, 276, 42619–42631.
16. Bloch, Konrad, "Summing up," *Annual Reviews of Biochemistry* 1987, 56, 1–9.
17. Kennedy, Eugene P., "Konrad Bloch," *Proceedings of the American Philosophical Society*, 2003, 147(10), 67–72.

18. Steinberg, Daniel, "Thematic review series: The pathogenesis of atherosclerosis. An interpretive history of the cholesterol controversy, part III: Mechanistically defining the role of hyperlipidemia," *Journal of Lipid Research* 2005, 46(10), 2037–2051.
19. Bloch, Konrad, *Blondes in Venetian Paintings, the Nine-Banded Armadillo, and Other Essays in Biochemistry*. Yale University Press: New Haven, 1994.
20. Fitzgerald, Brian, "Framingham Heart Study director Daniel Levy's new book, *A Change of Heart*, a revolution in medicine: Interview with Daniel Levy," *BU Bridge*, April 29, 2005, 8(29).
21. Dawber, Thomas Royle, *The Framingham Study: The Epidemiology of Atherosclerotic Disease*, Harvard University Press: Cambridge, MA, 1980.
22. Brody, Jane E., "Heart diseases are persisting in study's second generation," *New York Times*, January 5, 1994.
23. Daniel, Levy; Brink, Susan, *A Change of Heart: How the Framingham Heart Study Helped Unravel the Mysteries of Cardiovascular Disease*, Alfred K. Knopf: New York, 2005, preface.
24. Kolata, Gina, "Cholesterol–heart disease link illuminated. New findings explain how blood cholesterol levels are controlled and how to lower them substantially in persons at high risk of heart disease," *Science* 1983, 221, 1164–1166.
25. After the concept of "LDL receptor" was advanced by Brown and Goldstein, Yosio Watanabe collaborated with his biochemistry colleagues at the Kobe University to look for LDL receptors on cultured skin cells from those rabbits. They found that the rabbits, like FH humans, lack functional LDL receptors.
26. Roberts, Royston M., *Serendipity, Accidental Discoveries in Science*, John Wiley and Sons: Hoboken, NJ, 1989, p. 137.
27. Brown, Michael S.; Goldstein, Joseph L., "A receptor-mediated pathway for cholesterol homeostasis," *Science* 1986, 232, 34–47; Goldstein, Joseph L.; Brown, Michael S., "The cholesterol quartet," *Science* 2001, 292, 1310–1312.
28. Goldstein, Joseph L., "The Nobel Banquet speech," December 10, 1985. Available at: http://nobelprize.org/nobel_prizes/medicine/laureates/1985/goldstein-speech.html.
29. Rabin, Roni, "Value of cholesterol targets is disputed," *New York Times*, October 17, 2006.

CHAPTER 2

1. Altschul, Rudolf; Hoffer, Abram; Stephen, J. D., "Influence of nicotinic acid on serum cholesterol in man," *Archives of Biochemistry and Biophysics* 1955, 54, 558–559.
2. Parsons, W. B.; Achor, R. W. P.; Berge, K. G.; McKenzie, B. F.; Barker, N. W., "Changes in blood lipids following prolonged administration of large doses of nicotinic acid to persons with hypercholesterolemia," *Mayo Clinic Proceedings* 1956, 31, 377.
3. Lorenzen, A.; Stannek, C.; Lang, H.; et al., "Characterization of a G protein-coupled receptor for nicotinic acid," *Journal of Biological Chemistry* 2001, 59, 349–357.

4. Karpe, F.; Frayn, K. N., "The nicotinic acid receptor—a new mechanism for an old drug," *Lancet* 2004, 363, 1892–1894.
5. Carlson, L. A., "Nicotinic acid: The broad-spectrum lipid drug. A 50th anniversary review," *Journal of Internal Medicine* 2005, 258(2), 94–114; Pierson, Randell, "Merck cites new cholesterol drugs, boosts cost cuts," *Reuters Financial News*, December 15, 2005.
6. Hashim, Sami A.; van Itallie, Theodore B., "Cholestyramine resin therapy for hypercholesteremia. Clinical and metabolic studies," *JAMA* 1965, 192(4), 289–293.
7. Landau, Ralph; Achilladelis, Basil; Scriabine, Alexander (eds.), *Pharmaceutical Innovation, Revolutionizing Human Health*, Chemical Heritage Foundation: Philadelphia, 1999, p. 207.
8. "Company news; Upjohn to market cholesterol-lowering tablet," *Time*, July 23, 1994.
9. Thorp, J. M.; Waring, W. S., "Modification of metabolism and distribution of lipids by ethyl chlorophenoxyisobutyrate," *Nature* 1962, 194, 948–949; Sneader, Walter, *Drug Prototypes and Their Exploitation*, John Wiley and Sons: New York, 1996, p. 684.
10. Sneader, Walter, *Drug Discovery: A History*, John Wiley and Sons: New York, 2004, p. 275.
11. Palopoli, Frank P., "Basic research leading to MER-29," *Progress in Cardiovascular Diseases* 1960, 2, 489–491.
12. Gold, Mark S.; Boyette, Michael, *Wonder Drugs, How They Work*, Pocket Books: New York, 1982, p. 81.
13. Fine, Ralph Adam, *The Great Drug Deception: The Shocking Story of MER/29 and the Folks Who Gave You Thalidomide*, Stein and Day: New York, 1972; Knightley, Phillip; Evans, Harold; Potter Elaine; Wallace, Marjorie, *Suffer the Children: The Story of Thalidomide*, Viking Press: New York, 1979.
14. "Triparanol Side Effects," *Time* April 3, 1964.
15. Landers, Peter, "Stalking cholesterol: How one scientist intrigued by molds found first statin," *Wall Street Journal*, January 9, 2006.
16. Endo, Akira, "The origin of the statins," *Atherosclerosis Supplements* 2004, 5(3), 125–130.
17. Endo, Akira, "The discovery and development of HMG-CoA reductase inhibitors," *Journal of Lipid Research* 1992, 33(11), 1569–1582. Reprinted as: Endo, A. "The discovery and development of HMG-CoA reductase inhibitors," *Atherosclerosis Supplements* 2004, 5(3), 67–80.
18. Endo, A., "Mevastatin (ML-236B) and related compounds as potential cholesterol-lowering agents that inhibit HMG-CoA reductase," *Journal of Medicinal Chemistry* 1985, 28(4), 401–405; Endo A.; Kuroda M.; Tanzawa K., "Competitive inhibition of 3-hydroxy-3-methylglutaryl coenzyme A reductase by ML-236A and ML-236B fungal metabolites, having hypocholesterolemic activity. 1976," *Atherosclerosis Supplements* 2004, 5(3), 39–42.
19. Petersen, Melody, "Celebrities join Bristol-Myers campaign," *New York Times*, November 13, 2001.
20. Brown, Michael S.; Goldstein, Joseph L., "A tribute to Akira Endo, discoverer of a 'penicillin' for cholesterol," *Atherosclerosis Supplements* 2004, 5(3), 13–16.

21. Endo, Akira, "I finally tested a statin on myself!!" *Atherosclerosis Supplements* 2004, 5(3), 81.

CHAPTER 3

1. Hawthorne, Fran, *The Merck Druggernaut: The Inside Story of a Pharmaceutical Giant*, John Wiley and Sons: Hoboken, NJ, 2003, pp. 20–23.
2. Gortler, Leon, "Merck in America: The first 70 years from fine chemicals to pharmaceutical giant," *Bulletin for the History of Chemistry* 2000, 25(1), 1–9.
3. Lynn, Matthew, *The Billion-Dollar Battle: Merck v. Glaxo*, Heinemann: London, 1991, p. 80.
4. Andrews, George; Solomon, David (eds.), *The Coca Leaf and Cocaine Papers*, Harcourt Brace Jovanovich: New York, NY, 1975.
5. Cloud, John, "The lure of ecstasy," *Time*, June 5, 2000.
6. "What the doctor ordered," *Time*, August 18, 1952.
7. Lax, Eric, *The Mold in Dr. Florey's Coat*, John Macrae: New York, 2004.
8. Brown, Kevin, *Penicillin Man: Alexander Fleming and the Antibiotic Revolution*, Sutton: London, 2004.
9. Sarett, Lewis H., "Clyde Max Tishler, October 30, 1906–March 18, 1989," *Biographical Memoirs, National Academy of Sciences and Arts* 1989, 353–369.
10. Vagelos, P. Roy; Galambo, Louis, *Values and Visions: A Merck Century*, Merck: Whitehouse Station, NJ, 1991.
11. Sheehan, John, C., *Enchanted Ring: The Untold Story of Penicillin*, MIT Press, Cambridge, MA, 1982.
12. Patchett, Arthur A., "Lewis Hastings Sarett, December 22, 1917–November 29, 1999," *Biographical Memoirs, National Academy of Sciences and Arts* 2002, 81, 1–17.
13. Sarett, Lewis H., "Perhaps this job isn't right for you," *R&D Innovator* 1995, 4(5), 155.
14. Hirschmann, Ralph, "The cortisone era; aspects of its impact. Some contributions of the Merck Laboratories," *Steroids* 1992, 57, 579–592.
15. Beyer, K. H., "Discovery of the thiazide: Where biology and chemistry meet," *Perspectives in Biology and Medicine* 1977, 20, 410–420.
16. deStevens, George, "My odyssey in drug discovery," *Journal of Medicinal Chemistry* 1991, 34(9), 2665–2670.
17. Patchett, Arthur A., "Enalapril and lisinopril," in Lednicer, Daniel (ed.), *Chronicles of Drug Discovery*, American Chemical Society: Washington, DC, 1993, pp. 125–162; Patchett, "Excursions in drug discovery," *Journal of Medicinal Chemistry* 1993, 36(15), 2051–2058; Patchett, "Natural products and design: Interrelated approaches in drug discovery," *Journal of Medicinal Chemistry* 2002, 45(26), 5609–5616.
18. Vagelos, P. Roy; Galambo, Louis, *Medicine, Science and Merck*, Cambridge University Press: Cambridge, 2004.
19. Vagelos, P. Roy, "Are prescription drug prices high?" *Science* 1991, 252, 1080–1084.

20. Endo, A., "The origin of the statins," *Atherosclerosis Supplements* 2004, 5(3), 125–130.
21. Landers, Peter, "Stalking cholesterol: How one scientist intrigued by molds found first statin," *Wall Street Journal*, January 9, 2006.
22. Alberts, Alfred W., "Discovery, biochemistry and biology of lovastatin," *American Journal of Cardiology* 1988, 62, 10J–15J.
23. Tobert, Jonathan A., "Case history: Lovastatin and beyond: The history of the HMG-CoA reductase inhibitors," *Nature Reviews Drug Discovery* 2003, 2(7), 517–526.
24. Brody, Jane E., "New type of drug for cholesterol approved and hailed as effective," *New York Times*, September 2, 1987.
25. Moore, Thomas J., *Heart Failure: A Searching Report on Modern Medicine at Its Best...and Its Worst*, Simon and Schuster: New York, 1989, pp. 27–95.
26. Stolberg, Sheryl Gay, "Judge rebuffs drug agency on effort to ban diet supplement," *New York Times*, June 17, 1998.
27. Couzin, Jennifer, "The brains behind blockbusters," *Science* 2005, 309, 728–730.
28. "Patents; cholesterol drug brings honors to 4," *New York Times,* April 16, 1988.
29. Altman, Lawrence K., "Clinton operation aims to restore blood flow," *New York Times*, September 4, 2004.
30. Mcfadden, Robert D.; Altman, Lawrence K., "Clinton is given bypass surgery for 4 arteries," *New York Times*, September 7, 2004.

CHAPTER 4

1. Hoefle, Milton L., "The early history of Parke-Davis and Company," *Bulletin for the History of Chemistry* 2000, 25(1), 28–34; Mahoney, Tom, *The Merchants of Life: An Account of the American Pharmaceutical Industry*, Harper: New York, 1959, p. 31.
2. Galambos, Louis; Stewell, Jane Eliot, *Network of Innovation: Vaccine Development at Merck, Sharp and Dohme, and Mulford*, Cambridge University Press: Cambridge, 1995, p. 10.
3. Sneader, Walter, "The discovery and synthesis of epinephrine," *Drug News and Perspectives* 2001, 14(8), 491–494.
4. Bowden, Mary Ellen, "The rush to adrenaline," *Chemical Heritage* 2003, 21(1), 12–40.
5. Friedlander, Walter J. "Putnam, Merritt, and the discovery of Dilantin," *Epilepsia* 1986, 27 (supplement 3), S1–S21.
6. Dreyfus, Jack, *The Story of a Remarkable Medicine*, Lantern Books: New York, 2003.
7. Sneader, Walter, *Drug Discovery: A History*, John Wiley and Sons: New York, 2005, pp. 302–303.
8. Landau, Ralph; Achilladelis, Basil; Scriabine, Alexander (eds.), *Pharmaceutical Innovation—Revolutionizing Human Health*, Chemical Heritage Press: Philadelphia, 1999, p. 172.
9. Asbell, Bernard, *The Pill, A Biography of the Drug That Changed the World*, Random House: New York, 1995, p. 65.

10. Marks, Lara V., *Sexual Chemistry, A History of the Contraceptive Pill*, Yale University Press: New Haven, 2001, p. 165.
11. Djerassi, C., *The Pill, Pygmy Chimps, and Degas' Horse*, Basic Books: New York, 1992, p. 60.
12. Thorp, J. M.; Waring, W. S., "Modification of metabolism and distribution of lipids by ethyl chlorophenoxyisobutyrate," *Nature* 1962, 194, 948–949.
13. Sneader, Walter, *Drug Discovery: The Evolution of Modern Medicine*, John Wiley and Sons: New York, 1985.
14. Freudenheim, Milt, "Business and health; cholesterol drugs' growth," *New York Times*, October 6, 1987.
15. Frick, M. H.; Elo, O.; Haapa, K.; Heinonen, O. P.; Heinsalmi, P.; Helo, P.; Huttunen, J. K.; Kaitaniemi, P.; Koskinen, P.; Manninen, V.; et al. "Helsinki Heart Study: Primary-prevention trial with gemfibrozil in middle-aged men with dyslipidemia. Safety of treatment, changes in risk factors, and incidence of coronary heart disease," *New England Journal of Medicine*, November 12, 1987.
16. Kolata, Gina, "Cholesterol-altering drug found to reduce risk of heart attack," *New York Times*, November 12, 1987.
17. F.D.A. backs Warner drug," *New York Times*, January 19, 1989.
18. "Lopid worry for Lambert," *New York Times*, February 28, 1992.
19. "Company news; Marion Merrell gets approval to make generic drugs," *New York Times*, June 8, 1995.
20. Informal discussion with Roger S. Newton, November 23, 2005.
21. Informal discussion with Catherine S. Sekerke, March 21, 2006.
22. Goldstein, Joseph L.; Basu, Sandip K.; Brown, Michael S., "Receptor-mediated endocytosis of low-density lipoprotein in cultured cells," *Methods in Enzymology* 1983, 98, 241–260.
23. Informal discussion with Bruce D. Roth, March 20, 2006; Roth, B. D., "The discovery and development of atorvastatin, a potent novel hypolipidemic agent," *Progress in Medicinal Chemistry* 2002, 40, 1–22.
24. Winslow, Ron, "The birth of a blockbuster: Lipitor's route out of the lab," *Wall Street Journal*, January 24, 2000.
25. Willard, Alvin K.; Novello, Frederick C.; Hoffmann, William F.; Cragoe, Edward J., Jr., (+)-(4R,6S)-(E)-6-[2-(4'-Fluoro-3,3',5-trimethyl-[1,1'-biphenyl]-2-yl)ethenyl]-3,4,5,6-tetrahydro-4-hydroxy-2H-pyran-2-one and a Pharmaceutical Composition Containing It. European Patent Application EP19810105013, 1983; Willard, Alvin K.; Novello, Frederick C.; Hoffman, William F.; Cragoe, Edward J., Jr., Substituted Pyranone Inhibitors of Cholesterol Synthesis. U.S. Patent 4,567,289, 1986; Stokker, G. E.; Alberts, A. W.; Anderson, P. S.; Cragoe, E. J., Jr.; Deana, A. A.; Gilfillan, J. L.; Hirshfield, J.; Holtz, W. J.; Hoffman, W. F.; Huff, J. W.; Lee, T. J.; Novello, Frederick C.; Prugh J. D.; Rooney, C. S.; Smith, R. L.; Willard, Alvin K., "3-Hydroxy-3-methylglutaryl-coenzyme A reductase inhibitors. 3. 7-(3,5-Disubstituted-[1,1'-biphenyl]-2-yl)-3,5-dihydroxy-6-hept-enoic

acids and their lactone derivatives," *Journal of Medicinal Chemistry* 1986, 29, 170–181.
26. Kathawala, Faizulla Gulamhusein, Analogs of Mevalolactone and Derivatives Thereof and Their Use as Pharmaceuticals. U.S. Patent 5,354,772, 1984; F. G. Kathawala, Stabilized Pharmaceutical Compositions Comprising an HMG-CoA Reductase Inhibitor Compound. U.S. Patent 4,739,073, 1988.
27. Kathawala, F. G., "HMG-CoA reductase inhibitors: An exciting development in the treatment of hyperlipoproteinemia," *Medicinal Research Reviews* 1991, 11(2), 121–146.
28. Roth, B. D.; Ortwine, D. F.; Hoefle, M. L.; Stratton, C. D.; Sliskovic, D. R.; Wilson, M. W.; Newton, R. S., "Inhibitors of cholesterol biosynthesis. Part I. Trans-6-[2-(1H-pyrrol-1-yl)ethyl]-4-hydroxy-pyran-2-ones, a novel series of HMG-CoA reductase inhibitors. 1. Effects of structural modifications at the two and five positions of the pyrrole nucleus," *Journal of Medicinal Chemistry* 1990, 33, 21–31.
29. Sigler, Robert E.; Dominick, Mark A.; McGuire, Edward J., "Subacute toxicity of a halogenated pyrrole hydroxymethylglutaryl-coenzyme A reductase inhibitor in Wistar rats," *Toxicological Pathology* 1992, 20(4), 595–602.
30. Informal discussion with Bob Sliskovic, August 24, 2006.
31. Sliskovic, D. R.; Roth, B. D.; Wilson, M. W.; Hoefle, M. L.; Newton, R. S., "Inhibitors of cholesterol biosynthesis. 2. 1,3,5-Trisubstituted [2-(tetrahydro-4-hydroxy-2-oxopyran-6-yl)ethyl]pyrazoles," *Journal of Medicinal Chemistry* 1990, 33, 31–38.
32. Wareing, James Richard, (Pyrazolylvinyl)mevalonic Acids (Sandoz A.-G., Switzerland). U.S. patent 4613610, 1986.
33. Roth, B. D., Blankley, C. J., Chucholowski, A. W., Ferguson, E., Hoefle, M. L., Ortwine, D. F., Newton, R. S., Sekerke, C. S., Sliskovic, D. R., Stratton, C. D., Wilson, M. W., "Inhibitors of cholesterol biosynthesis. Part 3. Tetrahydro-4-pyran-2-one inhibitors of HMG-CoA reductase. 2. Effects of introducing substituents at positions three and four of the pyrrole nucleus," *Journal of Medicinal Chemistry* 1991, 34, 357–366.
34. Simons, John, "The $10 billion pill," *Fortune*, January 6, 2003.
35. The plight of the closing of the Ann Arbor laboratories is chronicled by Avery Johnson in "As drug industry struggles, chemists face layoff wave," *Wall Street Journal*, December 11, 2007.

CHAPTER 5

1. Informal discussion with Bruce D. Roth, March 20, 2006; Roth, B. D., "The discovery and development of atorvastatin, a potent novel hypolipidemic agent," *Progress in Medicinal Chemistry* 2002, 40, 1–22.
2. Li, Jie Jack, *Laughing Gas, Viagra, and Lipitor: The Human Stories behind the Drugs We Use*, Oxford University Press: New York, 2006, pp. 86–90.
3. Landau, Ralph; Achilladelis, Basil; Scriabine, Alexander (eds.), *Pharmaceutical Innovation—Revolutionizing Human Health*, Chemical Heritage Press: Philadelphia, 1999, pp. 186–188.

4. Hoefle, M. L.; Klutchko, S., "Substituted Acyl Derivatives of 1,2,3,4-tetrahydroisoquinoline-3-carboxylic Acids. U.S. Patent 4344949, 1982 (Warner-Lambert).
5. Informal discussion with Roger S. Newton, November 23, 2005; Simons, John, "The $10 billion pill," *Fortune*, January 6, 2003.
6. Shook, Robert L., *Miracle Medicines: Seven Lifesaving Drugs and the People Who Created Them*, Portfolio: The Woodlands, TX, 2007.
7. Winslow, Ron, "The birth of a blockbuster: Lipitor's route out of the lab," *Wall Street Journal*, January 24, 2000.
8. Brower, Philip L.; Butler, Donald E.; Deering, Carl F.; Le, Tung V.; Millar, Alan; Nanninga, Thomas N.; Roth, Bruce D., "The synthesis of (4R-cis)-1,1-dimethylethyl 6-cyanomethyl-2,2-dimethyl-1,3-dioxane-4-acetate, a key intermediate for the preparation of CI-981, a high potent, tissue selective inhibitor of HMG-CoA reductase," *Tetrahedron Letters* 1992, 33(17), 2279–2282; Baumann, Kelvin L.; Butler, Donald E.; Deering, Carl F.; Mennen, Kenneth E.; Millar, Alan; Nanninga, Thomas N.; Palmer, Charles W.; Roth, Bruce D., "The convergent synthesis of CI-981, an optically active, highly potent, tissue-selective inhibitor of HMG-CoA reductase," *Tetrahedron Letters* 1992, 33(17), 2283–2284.
9. Jones, James H., *Bad Blood: The Tuskegee Syphilis Experiment*, Free Press: New York, 1993.
10. Reverby, Susan M. (ed.), *Tuskegee's Truths: Rethinking the Tuskegee Syphilis Study*, University of North Carolina Press: Chapel Hill, 2000, p. 575.
11. Rosenthal, Elisabeth, "British rethinking rules after ill-fated drug trial," *International Herald Tribune*, April 8, 2006.
12. Informal discussion with David Canter, October 25, 2006.
13. Nawrocki, J. W.; Weiss, S. R.; Davidson, M. H.; Sprecher, D. L.; Schwartz, S. L.; Lupien, P. J.; Jones, P. H.; Haber, H. E.; Black, D. M., "Reduction of LDL cholesterol by 25% to 60% in patients with primary hypercholesterolemia by atorvastatin, a new HMG-CoA reductase inhibitor," *Arteriosclerosis, Thrombosis, and Vascular Biology* 1995, 15(5), 678–682.
14. Hawthorne, Fran, *Inside the FDA: The Business and Politics behind the Drugs We Take and the Food We Eat*, John Wiley and Sons: Hoboken, NJ, 2005, pp. 92–93.
15. Bakker-Arkema, Rebecca G.; Davidson, Michael H.; Goldstein, Robert J.; Davignon, Jean; Isaacsohn, Jonathan L.; Weiss, Stuart R.; Keilson, Leonard M.; Brown, W. Virgil; Miller, Valery T.; et al., "Efficacy and safety of a new HMG-CoA reductase inhibitor, atorvastatin, in patients with hypertriglyceridemia," *JAMA* 1996, 275(2), 128–133; Bakker-Akerma, Rebecca G.; Best, James; Fayyad, Rana; Heinonen, Therese M.; Marais, A. David; Nawrocki, James W.; Black, Donald M., "A brief review paper of the efficacy and safety of atorvastatin in early clinical trials," *Atherosclerosis* 1997, 131(1), 17–23.
16. Black, Donald M.; Bakker-Arkema, Rebecca G.; Nawrocki, James W., "An overview of the clinical safety profile of atorvastatin (Lipitor), a new HMG-CoA reductase inhibitor," *Archives of Internal Medicine* 1998, 158(6), 577–584.

17. Bakker-Arkema, R. G.; Nawrocki, J. W.; Black, D. M., "Safety profile of atorvastatin-treated patients with low LDL-cholesterol levels," *Atherosclerosis* 2000, 149(1), 123–129.
18. Informal discussion with Mi Dong, January 5, 2007.

CHAPTER 6

1. Mohoney, John Thomas, *The Merchants of Life: An Account of the American Pharmaceutical Industry*, Harper: New York, 1959, pp. 237–252.
2. Tanner, Ogden, *25 Years of Innovation: The Story of Pfizer Central Research*, Greenwich: Lyme, CT, 1996.
3. "Penicillin grows in Brooklyn," *Time*, May 20, 1946.
4. Li, Jie Jack, *Laughing Gas, Viagra, and Lipitor: The Human Stories behind the Drugs We Use*, Oxford University Press: New York, 2006, p. 59; Otfinoski, Steven, *Alexander Fleming: Conquering Disease with Penicillin*, Facts on File: New York, 1993; Lax, Eric, *The Mold in Dr. Florey's Coat*, John Macrae: New York, 2004; Jacobs, Francine, *Breakthrough: The True Story of Penicillin*, New York: Dodd, Mead, 1985; Hobby, Gladys L., *Penicillin: Meeting the Challenge*, Yale University Press: New Haven, 1985.
5. Li, *Laughing Gas*, pp. 67–69.
6. Brunner, R., "Terramycin," *Scientia Pharmaceutica*, 1954, 22, 37–43; Stempel, Edward, "Strides in the development of the 'tetracyclines,'" *American Journal of Pharmacy and the Sciences Supporting Public Health* 1962, 134, 114–132.
7. "The line was very busy," *Time*, December 19, 1955.
8. Terrett, Nicholas K.; Bell, Andrew S.; Brown, David; Ellis, Peter, "Sildenafil (Viagra), a potent and selective inhibitor of type 5 cGMP phosphodiesterase with utility for the treatment of male erectile dysfunction," *Bioorganic and Medicinal Chemistry Letters* 1996, 6(15), 1819–1824.
9. Katzenstein, Larry, *Viagra: The Remarkable Story of the Discovery and Launch*, Medical Information Press: New York, 2001; Osterloh, Ian H., "The discovery and development of Viagra (sildenafil citrate)," in Dunzendorfer, Udo (ed.), *Sildenafil*, Birkhäuser: Basel, Switzerland, 2004, pp. 1–13.
10. "A record pace for Viagra sales," *New York Times*, July 7, 1998.
11. Koe, B. Kenneth; Harbert, Charles A.; Sarges, Reinhard; Weissman, Albert; Welch, Willard M., "Discovery of sertraline (Zoloft)," in Abstracts of Papers, 231st ACS National Meeting, Atlanta, GA, March 26–30 (2006), p. MEDI-185; Quallich, George J., "Development of the commercial process for Zoloft/sertraline," *Chirality* 2005, 17(supplement), S120–S126.
12. Mullin, Rick, "ACS award for team innovation," *Chemical and Engineering News* 2006, 84(5), 51–52.
13. "The discovery of fluconazole," *Pharmaceutical News* 1995, 2(4), 9–12.
14. Richardson, Ken, "The discovery of fluconazole," *Contemporary Organic Synthesis* 1996, 3(2), 125–132; Richardson, K.; Cooper, K.; Marriot, M. S.; Tarbit, M. H.; Troke, P. F.; Whittle, P. J., "Discovery of fluconazole, a novel antifungal agent," *Reviews of Infectious Diseases* 1990, 12(supplement 3), S267–S271.

15. Petersen, Melody, "What's black and white and sells medicine?" *New York Times*, August 27, 2000.
16. Reidy, Jamie, *Hard Sell: The Evolution of a Viagra Salesman*, Andrew McMeel: Kansas City, 2005.
17. Berenson, Alex, "Mixed reviews for 2 of Pfizer's top drugs," *New York Times*, March 9, 2005.
18. "Company news; Pfizer wins appeal in court battle over drug," *New York Times*, February 28, 2004.
19. Morrow, David J.; Holson, Laura M., "Warner-Lambert gets Pfizer offer for $82.4 billion," *New York Times*, November 6, 1999.
20. Morrow, David J., "A slightly kinder and gentler era for hostile takeovers," *New York Times*, November 16, 1999.
21. Holson, Laura M.; Petersen, Melody, "A consumer products giant will most likely stay with what it knows," *New York Times*, January 25, 2000.
22. Cannon, C. P.; Braunwald, E.; McCabe, C. H.; et al., "Intensive versus moderate lipid lowering with statins after acute coronary syndromes," *JAMA* 2004, 350, 15.
23. Kolata, Gina, "My Drug Study Sounds Catchier Than Yours," *New York Times*, March 7, 2004.
24. Kolata, Gina, "New conclusions on cholesterol," *New York Times*, March 9, 2004.
25. de Lemos, James A.; Blazing, Michael A.; Wiviott, Stephen D.; et al., "Early intensive vs a delayed conservative simvastatin strategy in patients with acute coronary syndromes. Phase Z of the A to Z trial," *JAMA* 2004, 292, 1307–1316.
26. Tramboo, Sabia, "Battle royale between Ranbaxy and Pfizer over Lipitor," IPfrontline.com, August 8, 2006.
27. Owens, Joanna, "More court rulings safeguard Lipitor," *Nature Reviews Drug Discovery* 2006, 5(2), 96.
28. Bloomberg News, "Court invalidates a patent that Pfizer holds for Lipitor," August 3, 2006.

CHAPTER 7

1. Angerbauer, Rolf; Fey, Peter; Huebsch, Walter; Philipps, Thomas; Bischoff, Hilmar; Petzinna, Dieter; Schmidt, Delf; Thomas, Guenter, Preparation of 7-(4-aryl-3-pyridyl)-3,5-dihydroxy-6-heptenoates and Analogs as Hypocholesteremics. U.S. Patent 5032602, 1989.
2. Petersen, Melody; Berenson, Alex, "Papers indicate that Bayer knew of dangers of its cholesterol drug," *New York Times*, February 22, 2003.
3. "Warning questioned after a drug kills 31," *New York Times*, August 21, 2001.
4. Pogson, G. W.; Kindred, L. H.; Carper, B. G., "Rhabdomyolysis and renal failure associated with cerivastatin-gemfibrozil combination therapy," *American Journal of Cardiology* 1999, 83(7), 1146; see also Bermingham, R. P.; Whitsitt, T. B.; Smart, M. L.; Nowak, D. P.; Scalley, R. D., "Rhabdomyolysis in a patient receiving the combination

of cerivastatin and gemfibrozil," *American Journal of Health-System Pharmacy* 2000, 57(5), 461–464.

5. Normally, a drug–drug interaction occurs when the two drugs use the same liver enzyme for their metabolism, which eliminates them from the body. In this case, Lopid is a P450 2C8 inhibitor and an inhibitor of statin glucuronidation. In essence, not only did Lopid occupy the liver enzyme that metabolized Baycol, but it also prevented Baycol from being eliminated by forming the glucuronide. Since Baycol was not sufficiently metabolized and eliminated, the body began to accumulate Baycol, which would cause more serious side effects, most notably, rhabdomyolysis. See Touw, D. J.; Schalekamp, T.; van der Kuy, A.; van Loenen, A. C., "All statins are equal, but. A comparison of HMG-CoA reductase inhibitors," *Pharmaceutisch Weekblad* 2000, 135(36), 1338–1344.

6. Rodriguez, M. L.; Mora, C.; Navarro, J. F., "Cerivastatin-induced rhabdomyolysis," *Annals of Internal Medicine* 2000, 132(7), 598.

7. Ozdemir, O.; Boran, M.; Gokce, V.; Uzun, Y.; Kocak, B.; Korkmaz, S., "A case with severe rhabdomyolysis and renal failure associated with cerivastatin-gemfibrozil combination therapy—a case report," *Angiology* 2000, 51(8), 695–697.

8. "Withdrawal of drug, world business briefing, Asia: Japan," *New York Times*, August 24, 2001.

9. Greteman, Blaine; Zwick, Steve, "Bayer's bitter pill, drugs giant is forced to swallow hard as its stock market value slides," *Time Europe*, March 10, 2003.

10. Davidson, Michael H., "Rosuvastatin: A highly efficacious statin for the treatment of dyslipidemia," *Expert Opinion on Investigational Drugs* 2002, 11(1), 125–141; Rosenson, Robert S., "Rosuvastatin: A new inhibitor of HMG-CoA reductase for the treatment of dyslipidemia," *Expert Review of Cardiovascular Therapy* 2003, 1(4), 495–505.

11. "World business briefing, Europe: Britain: Cholesterol drug setback," *New York Times*, July 24, 2002.

12. "Company news; AstraZeneca cholesterol drug wins FDA approval," *New York Times*, August 13, 2003.

13. Horton, Richard, "The statin wars: Why AstraZeneca must retreat [editorial]," *Lancet* 2003, 362, 1341.

14. McKillop, Tom, "Astra's response," *Lancet* 2003, 362, 1498.

15. Wolfe, Sidney M., "Dangers of rosuvastatin identified before and after FDA approval," *Lancet* 2004, 263, 2189–2190; Harris, Gardiner, "Drug industry's longtime critic says 'I told you so,'" *New York Times*, February 15, 2005.

16. Tuler, David, "Seeking a fuller picture of statins," *New York Times*, July 20, 2004.

17. Gorman, Christine, "How safe are they? In a rare move, an FDA official openly attacks five drugs. Should you worry?" *Time*, November 21, 2004.

18. Altman, Lawrence K.; Saul, Stephanie, "Mixed safety results on cholesterol drug," *New York Times*, May 24, 2005.

19. Hirschler, Ben, "AstraZeneca's Crestor may get lift from new data," *Reuters*, March 8, 2006.

20. Reuters, "A drug is found to reduce plaque in arteries," March 14, 2006.
21. Rosenblum, Stuart B., "Chapter 13. Cholesterol absorption inhibitors: Ezetimibe (Zetia)," in Johnson, Douglas S.; Li, Jie Jack (eds.), *The Art of Drug Synthesis*, John Wiley and Sons: Hoboken, NJ, 2007.
22. Andrews, Michelle, "Money and medicine; a case for older (less expensive) drugs," *New York Times*, November 17, 2002.
23. Burnett, D. A.; Caplan, M. A.; Davis, H. R., Jr.; Burrier, R. E.; Clader, J. W., "2-Azetidinones as inhibitors of cholesterol absorption," *Journal of Medicinal Chemistry* 1994, 37, 1733–1736.
24. Clader, John W., "The discovery of ezetimibe: A view from outside the receptor," *Journal of Medicinal Chemistry* 2004, 47(1), 1–9.
25. "Cholesterol drug wins approval," *New York Times*, October 26, 2002.
26. "FDA approves cholesterol drug," *New York Times*, July 24, 2004.
27. Bloomberg News, "Cholesterol pill sales bolster 2 drug makers," October 21, 2006.
28. Winslow, Ron; Rubenstein, Sarah, "Study deals setback to cholesterol drug," *Wall Street Journal*, January 15, 2008.
29. Gordon, T.; et al., "High density lipoprotein as protective factor against coronary heart disease: The Framingham Study," *American Journal of Medicine* 1977, 62(5), 707–714.
30. Hirano, K.; Yamashita, S.; Nakajima, N.; Arai, T.; Maruyama, T.; Yoshida, Y.; Ishigami, M.; Sakai, N.; Kameda-Takemura, K.; Matsuzawa, Y., "Genetic cholesteryl ester transfer protein deficiency is extremely frequent in the Omagari area of Japan. Marked hyperalphalipoproteinemia caused by CETP gene mutation is not associated with longevity," *Arteriosclerosis, Thrombosis, and Vascular Biology* 1997, 17(6), 1053–1059.
31. Inazu, A.; Brown, M. L.; Hesler, C. B.; Agellon, L. B.; Koizumi, J.; Takata, K.; Maruhama, Y.; Mabuchi, H.; Tall, A. R., "Increased high-density lipoprotein levels caused by a common cholesteryl-ester transfer protein gene mutation," *New England Journal of Medicine* 1990, 323(18), 1234–1238.
32. Berenson, Alex, "A Pfizer scientist sees research dividends ahead," *New York Times*, July 18, 2006.
33. Ruggeri, Roger B.; Wester, Ronald T.; Magnus-Aryitey, George, "Discovery of CP-529,414: A potent synthetic inhibitor of human plasma cholesteryl ester transfer protein (CETP)," In Abstracts of Papers, 225th ACS National Meeting, New Orleans, LA, March 23–27, 2003.
34. Harper, Matthew; Langreth, Robert, "Pfizer's warning signs," *Forbes*, December 8, 2006.
35. "Test drug said to increase good cholesterol," *New York Times*, April 8, 2004.
36. Berenson, Alex, "End of drug trial is a big loss for Pfizer," *New York Times*, December 4, 2006.
37. Story, Louise; Saul, Stephanie, "Scrutiny of other heart drugs could grow after failed trial," *New York Times*, December 4, 2006.

38. Berenson, Alex; Pollack, Andrew, "Pfizer shares plummet on loss of a promising heart drug," *New York Times*, December 5, 2006.
39. Informal discussion with Catherine S. Sekerke, March 21, 2006.
40. Informal discussion with Roger S. Newton, November 23, 2005.
41. Lemonick, Michael D., "Drano for the heart," *Time*, November 9, 2003.
42. Watson, Karol E., "ApoA-I Milano," *Current Opinion in Cardiovascular, Pulmonary and Renal Investigational Drugs* 2000, 2(3), 263–265.

CHAPTER 8

1. Vagelos, P. Roy, "Are prescription drug prices high?" *Science* 1991, 252, 1080–1084; Vagelos, "Social benefits of a successful biomedical research company," *Proceedings of the American Philosophical Society* 2001, 145(4), 575–578.
2. Law, L.; Rudnicka, A. R., "Statin safety: A systematic review," *American Journal of Cardiology* 2006, 97(supplement): 52C–60C.
3. Winslow, Ron, "Studies points to drop in heart attacks, cholesterol—evidence underlines value of preventive strategies; overcoming obesity, diabetes," *Wall Street Journal*, October 12, 2005.
4. Nocera, Joe, "The dangers of swinging for the fences," *New York Times*, January 27, 2007.
5. Tuler, David, "Seeking a fuller picture of statins," *New York Times*, July 20, 2004.

INDEX

1,4-dihydropyridine, 139
2-azetidinones, 156
2',4'-diaminoazobenzene-
  4-sulfonamide, 109
3-hydroxy-3-methylglutaryl coenzyme A
  (HMG-CoA). See HMG-CoA
3,4-dibromopyrrole, 91
3,4-dichloropyrrole, 91
3,4-methylene-dioxy-methamphetamine
  (MDMA). See MDMA
4S study. See Scandinavian Simvastatin
  Survival Study

Abbott Laboratories, 33, 50, 74, 153, 159, 174
Abel, John Jacob, 72–73
ACAT. See acyl CoA: cholesterol
  acyltransferase
Accupril (quinapril), 102–115, 173
Accutane, 153
ACE inhibitors, 55–56, 69, 89, 102–103
acetanilide, 47
acetic acid, 16–19, 43
acetic acid → squalene → cholesterol
  cascade, 19, 43
acetyl CoA, 44
acipimox, 29
active pharmaceutical ingredient, 32, 93, 108
acyl CoA: cholesterol acyltransferase
  (ACAT), 155–156
Adams, Roger, 10, 76
Adams' catalyst, 10
adipocytes, 29

adrenal cortex deficiency, 4
adrenal glands, 4, 72
Adrenalin, 73
adrenaline, 72–73, 78
Albers-Schonberg, George, 61, 67
Alberts, Alfred W., 57–60, 67
Aldrich, Thomas, 72–73
alkaloids, 45–47
Allen, Robert E., 34
Altschul, Rudolf, 28–29
Alzheimer's disease, 112, 167, 169
American Cyanamid Company, 50, 92, 128,
  130–131
American Home Products, 74, 141
amlodipine. See Norvasc
amphetamine, 46
anacetrapib, 161
analytical ultracentrifuge, 14, 20
Anderson, George A., 127
Anderson, Rudolph J., 17, 19
anesthetic, 46, 79
angel dust. See PCP
Angerbauer, Rolf, 150
angina, 3, 5, 27, 132, 133, 139, 143
angioplasty, 143, 168
animal model, 23, 39, 41, 74, 84, 101, 104,
  110, 113, 155, 156
Anitschkov, Nikolai N., 12–14, 41
Ann Arbor, Michigan, 71, 98, 99, 106, 107,
  110, 111, 115, 119, 122, 123, 165
antibiotics, 47, 53, 54, 76, 77, 128, 129, 130, 138
anticonvulsant, 75

antidepressant, 134, 140
antiflushing molecule, 30
antihistamine, 79
antisense technology, 155, 165
ApoA-1, 162, 164, 165
apolipoprotein B, 156
apoliprotein B-100, 165
ApoMilano, 161–166
arachidonic acid pathway, 30
arrhythmia, 3, 75
arterial degeneration, 27
Aspergillus niger, 126
Aspergillus terreus, 60
aspirin, 68, 150, 151, 169
ASTEROID clinical trial, 154, 155
asthma, 73, 153
AstraZeneca, 88, 97, 152, 153, 159, 173
asymmetric synthesis, 88, 90, 107, 108
atherosclerosis, 3, 11, 12, 13, 15, 29, 41, 81, 82, 84, 86
atorvastatin. *See* Lipitor
atorvastatin calcium, 97
atrial fibrillation, 3, 139
Atromid-S, 32, 33, 80, 173
atropine, 45
atypical antipsychotics, 142
Aureomycin, 128, 129, 130, 173
azithromycin, 132, 138, 142, 174

barbituric acid, 74
Barton, Sir Derek, 6, 9, 10
Bartz, Quentin, 77
Basford, Frederick, 34
Bassett, Angela, 42
Baycol (cerivastatin), 41, 44, 97, 125, 149–151, 153, 154, 169, 173
Bayer, 47, 97, 125, 133, 136, 139, 150, 151, 161, 173
Benadryl, 79
benzothiadiazine derivative, 55
beta-blocker, 69

Bextra, 153
bile acid, 4, 6, 26, 28, 31–32, 40, 51, 52, 59, 83
bile acid sequestrants, 26. *See also* resins
Biltz, Heinrich, 75, 76
bioavailability, 56, 94, 95, 97, 106, 128, 137, 139, 150
bioisosteres, 90
biosynthesis of cholesterol, 17–19, 37–38, 43, 58, 59, 61
biosynthesis of fatty acid, 57
birth control pill, 78
Bischoff, Hilmar, 150
bismuth salts, 47
Black, Donald M., 114, 115, 116, 119
Black, Sir James W., 132
black box, 78
black tongue, 28
Bloch, Clore, 18
Bloch, Konrad E., 16–19, 37, 44
Bloch, Peter, 18
blockbuster drugs, 30, 67, 69, 89, 131, 132, 133, 137, 140, 142, 145, 170, 171
Bodnar, Andrew G., 144
Bolshevik Party, 12
Braun, Manfred, 107
Bright, G. Michael, 138
Bristol, Jim, 98
Bristol-Myers Squibb Company, 31, 32, 41, 42, 55, 67, 68, 69, 71, 89, 96, 97, 101, 118, 121, 125, 143, 144, 163, 171, 173, 174
Brown, A. G., 39
Brown, Michael S., 22–25, 31, 38, 39, 42, 54, 57, 59, 64, 82, 83
Burkholder, Paul Rufus, 76, 77
Burns, Margaret, 10
Butler, Donald E., 106
bypass, 68, 143, 168

C-reactive protein (CRP), 156, 169
Caduet, 145, 148
calcium-channel blockers, 89

Canter, David, 113–119, 122
Capoten (captopril), 55, 56, 102, 103, 173
captopril. *See* Capoten
Carvey, Dana, 42
Castelli, William, 21
cataracts, 35, 64
CD-28, 111
Celebrex (celecoxib), 142
celecoxib. *See* Celebrex
cell surface membrane protein, 24
Celontin, 75
Centro de Investigación Básica España (CIBE), 59, 67
cetirizine, 132, 174
cerivastatin. *See* Baycol
cervical cancers, 69
CETP. *See* cholesterol ester transfer protein
CETP gene, 160
Chain, Ernst, 9, 48, 49, 127
Chalatov, S., 12
Chen, G. M., 75
Chen, Julie S., 60, 61
Chevreul, Michel E., 5
Chinese fermented red rice, 66
chiral synthesis, 95, 96, 107, 108
chloral hydrate, 47
chloramphenicol, 76–78
Chloromycetin, 76–78. *See also* chloramphenicol
chlorothiazide, 55
chlorotrianisene, 34
chlortetracycline, 128, 129, 173
cholanic acid, 6
cholesterine, 5
cholesterol ester transfer protein (CETP), 155, 159
cholesterol metabolism, 24, 39
cholesteryl-7-hydroxylase, 41
Cholestin, 66
cholestyramine, 25, 30–33, 36, 59, 83, 94, 101, 119, 174

Cholybar, 32
Christmasterone, 11
CI-957, 91–93
CI-971, 94–97
CI-981, 96–118, 122
Cialis, 133
Ciba-Geigy, 87, 103, 122, 174
CIBE. *See* Centro de Investigación Básica España
Cipro (ciprofloxacin), 151
ciprofloxacin. *See* Cipro
citric acid, 126, 127
citrinin, 38
Clader, John W., 157
Claritin (loratadine), 157
Clarke, Hans T., 17
clinical trials, 27, 34, 40–41, 53, 56, 58, 61–63, 69, 75, 79, 81, 82, 97, 98, 103–105, 108–115, 117, 121, 122, 125, 133, 135, 144, 150, 152, 155, 157, 162, 165
Clinton, Bill, 68, 69, 110, 149
clofibrate, 32–33
clofibric acid, 33
clotrimazole, 136
cocaine, 46
Coconut Grove Club, 51
codeine, 45
coenzyme Q10, 44
Colestid, 31
colestipol, 31
compactin, 39. *See also* Mevastatin
Compañía Española de la Penicilina y Antibióticos, 59
congestive heart failure, 3, 102, 115
coniine, 45
Conover, Lloyd H., 129, 130
constipation, 32
Controulis, John, 77
Cordaptive, 30, 159
Cori, Carl, 57

Cornforth, John W., 8–9, 16
Cornforth, Rita, 8
Coronary Drug Project, 27
coronary heart disease, 3, 4, 13, 15, 20, 23, 25, 26, 37, 65, 68, 69, 80, 81, 118, 125, 149, 159, 165, 167, 168, 170
Coronary Primary Prevention Trial (LRC-CPPT), 25
cortisol, 4
cortisone, 11, 48, 51–53, 73–74
Creger, Paul L., 33, 86
Crestor (rosuvastatin), 64, 88, 97, 149–159, 168, 170, 173. *See also* rosuvastatin
Creswell, Ronnie, 86, 92, 95, 104
Crooks, Harry M., 77
CRP. *See* C-reactive protein
Cuemid, 32
curves, the, 116, 117
Cushman, David, 102
cyclooxygenase metabolite, 30

dalvastatin, 101, 104, 105, 106, 122
Dawber, Thomas "Roy," 20, 21
de Vink, Lodewijk J. R., 123, 141
deep-tank fermentation, 128
degenerative vascular diseases, 29
depression, 17, 75, 132, 134, 135, 142, 170
desmosterol, 34, 35
deStevens, George, 55
deuterium-labeled acetate, 17
diabetes, 20, 21, 25, 33, 69, 140, 165, 169
diastereomers, 96
dichlorophenamide, 55
differentiation, 113
Diflucan, 135–137
diflunisal. *See* Dolobid
Dilantin (phenytoin), 74–76
diphenhydramine, 79
diphenylhydantoin, 74
discrepancy between human and animals, 113
Diuril. *See* chlorothiazide
Djerassi, Carl, 78, 79

Dolobid (diflunisal), 55
Douglas, Kirk, 42
DPP IV inhibitor, 69
Dramamine, 79
Dreyfus, Jack, 75
Dr. Reddy's Laboratories, 140
drug–drug interaction, 33, 151
Dubos, René J., 53, 76
Duffield, Parke & Company, 72
Duffield, Samuel P., 72
Duggar, Benjamin M., 128
Dyson-Perrin Laboratory, 8, 9

E. R. Squibb & Sons, 47, 49, 127, 131
early cholesterol drugs, 27–33
Ecstasy, 46
Effect of Combination Ezetimibe and High-Dose Simvastatin vs. Simvastatin Alone on the Atherosclerotic Process in Patients with Heterozygous Familial Hypercholesterolemia (ENHANCE) trial. *See* ENHANCE trial
EH rabbit model, 93
Ehrlich, John, 77
Ehrlich, Paul, 24, 108
elemental analysis, 5
Eli Lilly, 50, 78, 122, 128, 133, 134, 137, 174
Elvejhem, Conrad, 28
enalaprilat, 56
enantiomer, 88, 96, 97, 105, 107, 135, 147, 148
Endo, Akira, 26, 27, 36–43, 45, 59, 61, 83, 89, 98
ENHANCE trial, 157, 160, 162
Enos, William, 15
epilepsy, 74–75
erectile dysfunction, 133
Erhart, Charles, 126
erythromycin, 137
Esperion, 163, 164, 165
estrogen, 4, 14, 27, 28, 33, 34, 159

ETC-216, 163, 164, 165
extra-cellular molecules, 24
eye, 35, 48, 64, 77, 110, 114
ezetimibe. *See* Zetia

familial hypercholesterolemia (FH), 23, 29, 40, 117, 157
Farnan, Joseph J. Jr., 147, 148
farnesyl-diphosphate, 44
FDA. *See* Food and Drug Administration
Federal Food, Drug and Cosmetic Act, 67
fenofibrate. *See* Tricor
Ferguson, Erika, 83
Fermentation Products for Screening Project (FERPS), 59, 61, 64
FERPS. *See* Fermentation Products for Screening Project
FH. *See* familial hypercholesterolemia
fibrates, 26, 28, 32–33, 36, 40, 63, 80, 150, 151, 155, 165, 168, 169
FIH. *See* first in human
FIP. *See* first in patients
first in human (FIH), 108, 114
first in patients (FIP), 114
Fischer, Emil, 6, 24
Fischer, Hans, 16, 17
Fleckenstein, Albrecht, 139
Fleming, Alexander, 48, 49, 127
Florey, Howard, 9, 48, 49, 127
fluconazole, 132, 173
fluoxetine, 134
flushing, 30
fluvastatin, 41, 171, 174
Folkers, Karl, 59
Food and Drug Administration (FDA), 30, 31, 32, 33, 34, 35, 36, 40, 41, 45, 56, 62, 63, 64, 65, 66, 68, 69, 75, 76, 78, 79, 81, 96, 97, 103, 105, 110–119, 123, 125, 129, 133, 134, 135, 137, 138, 140, 142, 144, 150–154, 157, 158, 159
Framingham Heart Study, 19–22, 25, 159

Framingham Tuberculosis Demonstration Study, 19
Freedland, Richard C., 82
Freud, Sigmund, 46
fungi, 37, 38, 129, 135, 136
Fuson, Robert L., 31

G-protein–coupled receptor, 29
Gadsden, Henry, 58
gallstone, 5
Gardasil, 69
Gattermann, Ludwig, 10
gemfibrozil. *See* Lopid
Genentech Inc., 99
generic invention, 146
generic names/drugs, 33, 61, 97, 118, 121, 139, 173, 174
Geodon, 142
glaucoma, 168
GlaxoSmithKline, 45, 68, 142, 150, 153, 174
gluconic acid, 127
GMP. *See* Good Manufacturing Practice
Gofman, John W., 14–15, 20
Goldstein, Joseph L., 22–25, 31, 38, 39, 42, 54, 57, 59, 64, 82, 83
Good Manufacturing Practice (GMP), 76
Graham, David, 153
gramicidin, 77
grapefruit juice effect, 151
Gregg, Alan, 49
griseofulvin, 136
Grundy, Scott, 62
guinea pig, 12, 13, 23, 101, 108, 113

HDL (high-density lipoprotein), 14, 15, 20, 30, 33, 39, 63, 65, 80, 112, 154, 159–165
heart attack, 3, 5, 15, 20, 21, 23, 25, 26, 28, 33, 63, 65, 66, 69, 80, 81, 82, 102, 117, 138, 143, 144, 149, 154, 155, 159, 162, 164, 167, 168
hematuria, 153
halofenate, 58

INDEX 193

Harrington, Charles, 9
Hashim, Sami A., 31
Hassan, Fred, 158
Heatley, Norman, 49
Helsinki Study, 80, 81
hemi-calcium salt, 97
Hench, Philip S., 52–53, 73
hepatocytes, 41, 82
heterozygous familial hypercholesterolemia, 23, 157
hexahydronaphthalene, 87, 88, 90
high blood pressure, 64, 69. *See also* hypertension
high-density lipoprotein. *See* HDL
high-throughput assay, 33, 60, 160
histamine, 54, 79, 89
Hitler, Adolf, 7
Hitler's gift, 16
HMG-CoA (3-hydroxy-3-methylglutaryl coenzyme A), 24, 36, 37, 38, 39, 43, 44, 59, 60, 61, 62, 65, 81, 83, 84, 86, 87, 89, 91, 92, 93, 95, 107, 110, 156
Hodgkin, Dorothy Crowfoot, 7
Hoechst, 89, 103, 173
Hoefle, Milton L., 81, 86
Hoffer, Abram, 29
Hoffman, Carl H., 59, 61, 67
Hoffman, William F., 65
Holmes, Robert H., 15
homozygous familial hypercholesterolemia, 23, 117
hormone, 4, 7, 11, 18, 27, 28, 32, 47, 72, 73, 78, 82
Huber, Wilson, 79
Huff, Jesse, 59
human papillomavirus vaccine, 69
Huttunen, Jussi, 80
hydrastinine, 46
hypercholesterolemic, 27, 29
hypertension, 3, 20, 25, 55, 102, 132, 134, 139, 142, 161, 165, 170

ICI (Imperial Chemical Industries Ltd.), 32, 33, 34, 80, 122, 173
Ignatowski, A. I., 12
Illingworth, Roger, 62
ILLUMINATE trial, 161
imipramine, 134
Imperial Chemical Industries Ltd. (ICI). *See* ICI
indole, 55, 88
inflammation, 169
intestinal lymphoma, 41
intestinal tumors, 41, 61, 62. *See also* intestinal lymphoma
Investigational New Drug, 40, 103, 110
iodine, 47
ischemic heart disease, 27, 63
isotope tracer technique, 17
Itallie, Theodore B. van, 31
ivermectin. *See* Mectizan

Janus, 4
Japan Tobacco, 161
Johnson & Johnson, 79
Jordan, Beulah, 35
JTT-705, 161

Kamm, Oliver, 76, 77
Kannel, William B., 20, 21
Kathawala, Faizulla G., 87–89
Kelsey, Frances, 36
Kendall, Edward C., 52, 73
Kende, Andrew S., 85
Kennedy, Eugene, 18
ketoconazole, 136
Keys, Ancel, 15–16
Keys, Margaret, 16
kidney, 72, 109, 150, 152, 153, 154
Kiliani, Heindrich, 6
Kimball, Dale A., 66
King, William, 36
Klutchkow, Sylvester, 103
Kobe beef, 23

Koe, B. Kenneth, 135
Kohler, Elmer P., 10, 50
Kolata, Gina, 80, 144
Köller, Carl, 46
Korean War, 15
Kos, 159
K-ration, 15
Kraus, George A., 84, 85
Kuroda, Masao, 38

lactone, 41, 88, 93
Ladenburg, Alfred, 28
Langdon, Robert, 18
Langham, Gerald George, 77
lanosterol, 18
laropiprant, 30
LDL (low-density lipoprotein), 3, 13, 14, 15, 20, 22, 23, 24, 25, 26, 29, 30, 31, 32, 39, 43, 57, 63, 65, 68, 80, 82, 83, 97, 98, 99, 101, 112, 113, 115, 116, 117, 118, 143, 144, 145, 149, 152, 154, 155, 156, 157, 159, 160, 161, 162, 164, 165, 167, 168, 169, 170
LDL receptor, 22–26, 31, 57, 82
Lederle laboratories, 49, 92, 128, 130, 173
Lescol, 41, 66, 71, 87–89, 97, 101, 104, 106, 110, 116, 121, 125, 150, 169, 170, 174
leukemia, 111
Levitra, 133
Liebig, Justus von, 72, 87
ligand and receptor concept, 24
Lindgren, Frank T., 14, 15
link between cholesterol and atherosclerosis, 13
lipids, 13, 17, 19, 20, 25, 30, 31, 32, 33, 37, 38, 40, 57, 58, 64, 68, 80, 82, 102, 112, 115, 119, 121, 159, 163
Lipitor (atorvastatin), 41, 44, 64, 71, 81, 82, 85, 86, 88, 89, 93, 94, 97, 98, 99, 101, 102, 105, 106, 108, 113–123, 125, 130, 132, 133, 140–148, 150, 153, 154, 157, 158, 160, 162, 163, 169, 170, 173

Lipitor basic patent, 146
Lipitor patent litigation, 145–148
Lipitor pure enantiomer patent, 147
Lipobay, 125
liver, 5, 8, 14, 23, 24, 34, 36, 39, 60, 64, 65, 76, 83, 90, 92, 96, 101, 112, 114, 136, 151, 156, 157, 169
lock and key hypothesis, 24
Long, Loren M., 75, 77
Lopez, Maria, 60
Lopid (gemfibrozil), 33, 36, 65, 80–81, 86, 105, 119, 150, 151, 169, 174
loratadine. *See* Claritin
Lorenze, Anna, 29
Lotrimin, 136
lovastatin. *See* Mevacor
low-density lipoprotein (LDL). *See* LDL
LRC-CPPT. *See* Coronary Primary Prevention Trial
LSD, 46
Lynen, Feodor, 16, 18, 19

Maanen, E. F. van, 36
macular degeneration, 168
Maddox, V. Harold, 79
Marion Merrell Dow Inc., 81
marker degradation, 78
Marker, Russell E., 78
Marmorston, Jessie, 28
Martin, Irwin, 116
Matsuzawa, Yuji, 160
Maxwell, Richard "Dick," 81, 82, 84
McGuire, Matthew M., 36
McKillop, Tom, 152
McKinnell, Henry (Hank) A., 142, 147, 175
MDMA, 46
Mead Johnson Laboratories, 31
Meadors, Gilcin, 20
mechanism-based, 43, 59, 62, 161
mechanism of action, 29, 43, 58, 103
Mectizan (ivermectin), 54

medicinal chemists, 55, 84, 86, 90, 106, 107, 150, 156, 157
menopause, 14
mental symptoms, 75
MER/29. *See* triparanol
Merck, 11, 30, 31, 35, 45–69, 73, 81, 82, 86, 87, 89, 95, 96, 101–104, 121, 122, 125, 127, 155–159, 161, 165, 167, 171
Merck, Friedrich Jacob, 45
Merck, George W., 46–48, 53, 54
Merck, Heinrich E., 45, 47
Meridia, 153
Merritt, H. Houston, 74
meta-analysis, 65
methsuximide, 75
me-too drug, 32, 40, 56, 102, 103
Mevacor (lovastatin), 33, 41, 43, 45, 46, 56–58, 61, 62, 64–68, 71, 82, 87–89, 93, 94, 97, 99, 101, 103–105, 110, 112, 116, 118, 120, 121, 125, 136, 150, 155, 167, 168, 170
mevalonate, 38, 43, 44, 62
mevalonic acid, 59–62, 64
Mevastatin, 27–28, 36–43, 45, 59–62, 65, 83, 87–92, 136, *see also* compactin
Mi Kyung Dong, 121
miconazole, 136
Miller, Ann, 50–51
Miller, C. A., 75
Milontin, 75
molasses, 127
molecular targets, 58, 132
molecule-based, 44, 161, 162
Monaghan, Richard L., 67
monkey, 13, 35, 39, 40, 156
Monsanto, 11
Moore, Thomas J., 65
morbidity and mortality, 26, 66
morphine, 45
motor symptoms, 75
multiple sclerosis, 168

Mulford Co., H. K., 72, 73
Murray, H. C., 11
muscle weakness, 40
mushroom, 37, 38, 135
myocardial infarction, 3, 145

naphthalene, 88
narcotics, 47
National Cholesterol Education Program (NCEP), 25, 26, 31
natural product, 59, 73, 87
Nazi(s), 7, 16, 131
NCEP. *See* National Cholesterol Education Program
NDA. *See* New Drug Application
Nelson, George, 84
neuron function, 22
neuropsychiatrist, 75
New Drug Application (NDA), 40, 63, 115, 121, 123, 125 *See*
Newton, Holmes, 81, 82
Newton, Roger S., 81–84, 86, 88, 94, 95, 96, 101, 103, 104, 105, 110, 114, 147, 162, 163, 164
niacin, 26, 28, 30, 36, 63, 159, 165, 168
Niaspan, 159
nicotinic acid, 26, 28–30, 36, 155, 159, 174. *See also* niacin
nicotinic acid receptor, 29
Niemann, Albert, 46
nifedipine, 139
Nirenberg, Marshall W., 22
Nissen, Steven E., 144, 158, 164
Nixon, Richard, 21
Nizoral, 136
NMR. *See* Nuclear Magnetic Resonance
Nobel Prize, 5, 6, 7, 8, 10, 14, 16, 19, 22, 24, 25, 37, 42, 53, 54, 57, 64, 73, 76, 132
nolo contendere, 36
"nonapprovable" letter, 30
norethindrone, 78, 79

Norvasc (amlopidine), 132, 138–140, 142, 145, 173
Novartis, 87, 97, 174
Novello, Frederick C., 55
Nuclear Magnetic Resonance (NMR), 6

Offspring Study, 21
Okazaki, H., 59
Ondetti, Miguel A., 102
Osterloh, Ian H., 133
ovaries, 4, 28
oxalic acid, 127
oxycholesterols, 29
oxytetracycline, 128, 129, 130, 131, 174

Paal-Knorr reaction, 107
Paal-Knorr synthesis of pyrroles, 107
Palopoli, Frank P., 34
Parke, Hervey C., 72
Parke-Davis, 33, 50, 55, 65, 71–97, 101–125, 128, 132, 140, 151, 162, 163
Patchett, Arthur A., 55, 56, 60, 102
Payne, Eugene, 77
PCP (phencyclidine hydrochloride *or* angel dust), 79
PDEs. *See* phosphodiesterases
PDM. *See* pharmacokinetics and drug metabolism
penicillin, 9, 38, 48–53, 76, 77, 89, 109, 127–129, 136
penicillin V, 50
Penicillium brevicompactum, 39
Penicillium notatum, 38, 48, 49
Perkin, William, Jr., 8
Perkin, William, Sr., 8
peroxisome proliferator-activated receptor-α (PPARα), 33, 155
Peterson, Durey H., 11
Pfizer, 45, 49, 50, 98, 99, 117, 122, 123, 125, 126–148, 153, 160, 161, 162, 163, 165, 173, 174

Pfizer, Charles, 49, 126, 127
Pharmacia, 131, 142, 164
pharmacokinetics and drug metabolism (PDM), 85, 94, 95, 97
Pharmanex Inc., 66
phase I clinical trials, 104, 105, 108, 110, 111, 112, 113, 114, 122, 133, 164
phase II clinical trials, 30, 113, 114, 115, 117, 120, 133, 160, 164, 165
phase III clinical trials, 30, 114, 115, 116, 118, 119, 133, 161, 165
phase IV clinical trials, 142, 143, 144
phencyclidine hydrochloride (PCP *or* angel dust), 79
phenobarbital, 74
phenytoin. *See* Dilantin
phosphodiesterases (PDEs), 133
Piasecki, Michelle, 85
PK/PD, 95
plant sterols, 32
plaque, 3, 4, 5, 6, 12, 13, 14, 20, 118, 144, 154, 155, 159, 160, 163, 164, 168
Plavix, 69
Pliva Pharmaceuticals, 137, 138
poison gas, 7
Popják, George, 16
potassium bromide, 74
potato sprouts, 8
Poulletier, François, 5
PPARα. *See* peroxisome proliferator-activated receptor-α
Pravacol (pravastatin), 41, 42, 67, 68, 71, 87, 89, 97, 101, 104, 110, 116, 118, 119, 120, 121, 125, 136, 143, 144, 150, 153, 154, 169, 170, 171, 174
pravastatin. *See* Pravacol
PRavastatin Or atorVastatin Evaluation and Infection Therapy (PROVE-IT), 118, 142–145
pregnanediol, 18
Premarin, 28

primates, 40
process chemists, 49, 50, 104, 106, 107, 108
Procter & Gamble, 141
product champion, 58, 61
progesterone, 4, 18, 78
prostaglandin, 30, 106, 107
prostaglandin D$_2$, 30
proteinuria, 153
PROVE-IT. *See* PRavastatin Or atorVastatin Evaluation and Infection Therapy
Prozac, 134
pteridines, 7
Public Citizen, 152, 153, 154
pulmonary hypertension, 134
Putnam, Tracy Jackson, 74
pyrazole, 92, 93
pyrimidine, 88, 92
pyrrole, 85, 90, 91, 92, 93, 94, 95, 107
pyrrolidine, 103
Pythium ultimum, 38

quadruple coronary bypass surgery, 68
quaternary ammonium ion exchange resin, 30
Questran, 32, 174
quinapril. *See* Accupril
quinine, 45

R-(+)-methylbenzylamine, 96
R1658, 161
rabbit, 12, 13, 15, 23, 29, 48, 93, 94, 160, 164
rabbit aortas, 12
Raetz, Christian R. H., 58
Ranbaxy Ltd., 140, 146–148
rational drug design, 102, 132, 155
rational drug discovery, 58
rat liver homogenate, 92
rat liver microsomes, 83
Rebstock, Mildred, 77
receptor, 22–26, 29, 30, 31, 33, 57, 64, 82, 89, 155

red blood cell, 4, 77
reductase inhibitor, 38, 39, 60, 61, 81, 84, 89, 93, 97, 105, 107
Reichstein, Tadeus, 74
Reinitzer, Friedrich, 5
resins, 26, 27, 28, 30–32, 36, 59, 63, 83, 101, 165, 168
Revatio, 134
Rezulin, 140
rhabdomyolysis, 150, 153, 169
rheumatoid arthritis, 11, 52, 73, 111
Rhône-Poulenc Rorer, 101, 104, 106, 122
Richardson-Merrell Inc., 34–36, 64, 81, 174
Rieveschl, George, 79
Rifkind, Basil, 80
risk-benefit profile, 150
risk factor, 3, 4, 15, 30, 21, 25, 37, 64, 138, 159, 162, 165, 167, 170
river blindness, 54
Robinson annulation reaction, 8
Robinson, Harry, 57, 58
Robinson, Sir Robert, 8–10, 16
Robson, John, 33
Roche, 50, 85, 87, 90, 103, 122, 153, 161, 162, 174
Rohm & Haas, 32, 89
Röntgen, Wilhelm C., 14
Rosenblum, Stuart B., 157
rosuvastatin. *See* Crestor
Roth, Bruce D., 84–86, 90, 91, 92, 93, 94, 96, 97, 99, 104, 107, 114, 126, 130, 147
Rothfield, Lawrence, 37
Rothrock, John, 60
Ruggeri, Roger B., 160
Rustein, David, 19
Ruzicka, Leopold, 16

salicylates, 47, 55
Sandoz Pharmaceuticals, 66, 71, 87–89, 93, 101, 104, 106, 121, 125

Sankyo Pharmaceuticals, 37–43, 45, 59, 61, 62, 101
Santonin, 126
sapogenin, 78
SAR. *See* structure-activity relationship
Sarett, Lewis H., 51–53, 56, 58
Scandinavian Simvastatin Survival Study (4S study), 65, 125
Schatz, Albert, 53, 54
Schering-Plough, 66, 155–158, 173, 174
schizophrenia, 29, 142
Schoenheimer, Rudolf, 17
Schonberg, Alexander, 33
Scott, Byron, 123
Seaborg, Glenn T., 14
Searle & Co., G. D., 74, 79, 122
sedative effect, 74
sedentary life style, 20, 25, 165
Sekerke, Catherine S., 83, 162
Seldin, Donald W., 22
selection invention, 147
selective serotonin reuptake inhibitor (SSRI), 134, 135, 140
semi-synthetic statin, 87
sensory symptoms, 75
serendipity, 27, 48, 132, 135, 156, 170
Serevent, 153
sertraline. *See* Zoloft
Seven Countries Study, 15–16
Sheehan, John C., 50, 52, 54
Shen, T. Y., 55
sildenafil. *See* Viagra
simvastatin. *See* Zocor
skin cells, 23
Sliskovic, Drago Robert, 86, 92, 93, 96, 97, 107
Smith, Nick, 139
Smith, Robert L., 65, 67, 99
smoking, 20, 25, 64, 86, 165
solanine, 8
Sonderhoff, R., 17

sorbose, 127
spleen, 29
Sprague-Dawly rats, 84
squalene, 18, 19, 43, 44
SSRI. *See* selective serotonin reuptake inhibitor
Stadtman, Earl R., 22, 28, 57
Stalin, Joseph, 12
Stallone, Sylvester, 42
Stamler, Jeremiah, 28
Steere, William C., Jr., 129, 142
Steinberg, Daniel, 62, 82, 83
Stephen, J. D., 29
Steptomyces venezuelae, 77
stereochemistry, 88
Stokker, Gerald E., 87
Streptomyces rimosus, 129
streptomycin, 48, 53–54, 76, 77, 128, 129
structure-activity relationship (SAR), 156
sucrose, 126
sulfa drugs, 47, 55, 109
sulfanilamide, 109
suprarenal glands, 72
Swiss albino mice, 49
Syntex, 78, 79, 122
syphilis, 108–110
systolic blood pressure, 161

tachycardia, 139
tadalafil, 133
Tagamet, 132
Takamine, Jokichi, 72, 73
Tchen, T. T., 18
technical defect of Lipitor patent, 148
Terrett, Nicholas, 133
testes, 4
testosterone, 4
tetracycline, 128, 130, 131
tetrahydroquinoline, 103
TGN1412, 111
thalidomide, 36, 41

Third Generation Study, 21
Thomas, H., 17
Thorp, J. M., 32, 80
thyroid, 13, 27, 73
thyroxine, 27, 63, 73, 78
time of application, 61
time of invention, 61
tioconazole, 136
Tishler, Max, 49–53, 56
tissue selectivity, 150
Tobert, Jonathan, 61, 62
Tofranil, 134
Topliss, John, 86, 90
Topliss Tree, 86
torcetrapib, 158–162, 171
total synthesis of cholesterol, 9
toxicology, 61, 63, 91, 93, 95, 106, 112, 137
Tricor (fenofibrate), 33, 159, 169, 174
triglyceride, 20, 30, 33, 101, 102, 112, 119, 151, 156, 159, 160, 163
triparanol (MER/29), 34–35, 43, 44, 64, 81, 174
triphenylethylene, 33
troglitazone, 140
Trosyl, 136
Turbostatin, 122
Tuskegee Syphilis Study, 109–110
type II diabetes, 33
typhus epidemic, 77
tyrothricin, 77

United States Adopted Names Council (USAN), 118
United States Patent and Trademark Office (USPTO), 138
Upjohn, 11, 29, 31, 32, 35, 50, 107, 122, 128, 131, 164, 173
USAN. *See* United States Adopted Names Council

USPTO. *See* United States Patent and Trademark Office

vaccine, 69, 72
Vagelos, P. Roy, 37, 54, 56–59, 62, 65, 67
vardenafil, 133
Vasotec, 56
veratrine, 45
Vfend (voriconazole), 142
Viagra (sildenafil), 132–134, 142, 174
Vick Chemical Company, 34
Vietnam War, 19
Vioxx, 69, 157
vitamin, 7, 27, 28, 30, 32, 47
vitamin B$_3$, 28
von Fürth, Otto, 72
voriconazole. *See* Vfend
Vytorin, 66, 69, 155–158, 170, 174

Wagner, Katherine "Coco," 82
Waksman, Selman Abraham, 53, 54, 76
Wallis, Everett S., 52
Wareing, James R., 93
Waring, W. S., 32, 80
Warner-Lambert, 32, 81, 85, 98, 104, 121, 122, 123, 125, 131, 132, 140, 141, 142
Watanabe, Yoshio, 23
water softener, 30–31
Watrous, Dick, 138
Weicker, Theodore, 46–47
Welch, Willard, 135
Werner, Harold, 36
Westheimer, Frank, 18
White, Paul Dudley, 21
white blood cells, 35, 102
Wieland, Heinrich O., 6–7
Wild, Anthony, 121
wild Mexican yams, 78
Willard, Alvin K., 65, 87, 90, 91, 96
William S. Merrell Company, 34

Willstätter, Richard M. 16, 17
Wilson, Jean, 62
Windaus, Adolf, 6–7, 18
Winter, Charles, 55
Wöhler, Friedrich, 46
Wolfe, Sidney M., 152
Woodruff, H. Boyd, 59, 61
Woodward, Arthur, 10
Woodward, Robert Burns, 8–11, 16, 18, 55, 87, 129
World War I, 7, 47, 131
World War II, 7, 15, 50, 131

X-ray, 7, 14, 39
X-ray diffraction, 7, 39

Yamamoto, Akira, 40

Zantac, 132
Zeller, James, 108
Zetia (ezetimibe), 66, 69, 155–158, 174
ziprasidone hydrochloride. *See* Geodon
Zithromax, 137–138
Zocor (simvastatin), 30, 41, 44, 56, 64–69, 71, 87–89, 99, 101, 104, 110, 116, 118, 119, 120, 121, 125, 143, 144, 149, 150, 153, 154, 155, 157, 158, 169, 170, 171
Zoloft (sertraline), 132–135, 142, 174
Zyrtec, 132, 142, 174